D0266220

79 308 199 0

The Eerie Silence

PAUL DAVIES

The Eerie Silence

Are We Alone in the Universe?

ALLEN LANE
an imprint of
PENGUIN BOOKS

Penguin Group (USA) Inc., 375 Hudson Street, New York, New York 10014, USA
Penguin Group (Canada), 90 Eglinton Avenue East, Suite 700, Toronto, Ontario, Canada M4P 2Y3
(a division of Pearson Penguin Canada Inc.)
Penguin Ireland, 25 St Stephen's Green, Dublin 2, Ireland (a division of Penguin Books Ltd)
Penguin Group (Australia), 250 Camberwell Road, Camberwell, Victoria 3124, Australia
(a division of Pearson Australia Group Pty Ltd)
Penguin Books India Pvt Ltd, 11 Community Centre, Panchsheel Park, New Delhi – 110 017, India
Penguin Group (NZ), 67 Apollo Drive, Rosedale, North Shore 0632, New Zealand
(a division of Pearson New Zealand Ltd)
Penguin Books (South Africa) (Pty) Ltd, 24 Sturdee Avenue, Rosebank, Johannesburg 2196, South Africa

Penguin Books Ltd, Registered Offices: 80 Strand, London WC2R 0RL, England

www.penguin.com

First published 2010
2

Set in 10.25/14 pt Postscript Adobe Sabon
Typeset by Ellipsis Books Limited, Glasgow
Printed in England by Clays Ltd, St Ives plc

ISBN: 978-1-846-14142-3

Contents

List of Illustrations

PLATES

FIGURES

Sometimes I think we're alone in the universe, and sometimes I think we're not. In either case the idea is quite staggering.

Arthur C. Clarke

Preface

In August 1931, Karl Jansky, a radio engineer working for Bell Telephone Laboratories in Holmdel, New Jersey, serendipitously made a major scientific discovery. Jansky had been assigned the task of investigating annoying radio static that interfered with transatlantic telephony. To check it, he built a simple antenna from metal struts, mounted on four car tyres so it could rotate, and proceeded to monitor radio noise from different directions. The output of the ramshackle instrument was a pen and ink recorder. Jansky was soon detecting thunderstorms, even far away, but he was puzzled by a background hiss that seemed to display a 24-hour cycle. Intrigued, he looked more closely and found the period to be 23 hours and 56 minutes, the duration known to astronomers as the sidereal day – the time it takes for Earth to rotate once relative to the distant stars (as opposed to the solar day, the time it takes to rotate relative to the sun). The sidereal periodicity implied that the radio source lay far out in space. Jansky eventually concluded that the radio static emanated from the Milky Way. Before he could follow up on it, however, he was assigned other duties by the company.

In this curiously low-key manner, an entire scientific discipline – radio astronomy – was born. No fanfare, no medals.[1] Further progress came, as so often in science, with war. The development of radar during the Second World War greatly boosted the power and fidelity of radio receivers, and in the immediate post-war years, physicists and astronomers spotted an opportunity. Using cheap left-over wartime equipment, they began to build the first proper radio telescopes, enormous dishes that enabled them to tune into the universe. About this time, in the 1950s, it dawned on some scientists that radio telescopes were powerful enough to communicate across interstellar distances, so that if there were any intelligent beings on other planets it would be possible for humans to receive their radio messages. On 19 September 1959 the respected scientific journal *Nature* published an article by two Cornell University physicists, Giuseppe Cocconi and Philip Morrison, entitled 'Searching for interstellar communications', in which the authors invited radio astronomers to look for radio messages coming from alien civilizations. Cocconi and Morrison conceded that their ideas were highly speculative, but concluded with the pertinent remark, 'The probability of success is difficult to estimate; but if we never search, the probability of success is zero.'[2] The following year, the challenge was taken up by a young astronomer, Frank Drake, to whom this book is dedicated. Drake used a radio telescope in West Virginia to begin searching for alien radio signals, and from his pioneering project the international research programme known as SETI was born. SETI stands for Search for Extraterrestrial Intelligence, and since the 1960s a heroic band of radio astronomers have been scouring the skies for any sign that we are not alone in the universe. In 2010, SETI will be officially fifty years old, which seems a good time to take stock. This book is a tribute to the dedication, professionalism and infectious optimism of SETI researchers in general, and to Frank Drake's courage and vision in particular.

The subject of SETI is speculative to a degree far beyond that of conventional science. It is wise to take any discussion of alien civilizations with a very large dose of salt. But retaining a robust scepticism need not prevent us from approaching SETI in a methodical and penetrating way, informed by the very best science we have. That is the spirit in which I have written this book. I have taken care to separate facts and theories in which we have some confidence, from reasonable but untested extrapolation, and from wilder speculation driven largely by ideas from science fiction.

I was still a high school student when SETI began, and although I was vaguely aware of it, my understanding of life beyond Earth was gleaned almost entirely from science fiction. Like many people, I learned more about SETI from the many television appearances of the charismatic scientist Carl Sagan, whose novel *Contact*, and the subsequent Hollywood movie based on it, convinced many people that SETI is a human adventure without parallel. In my later years, I came to know the key players quite well, many of whom now work at the SETI Institute in California. Much of what I have written about in this book stems from my long and fruitful association with them, especially Frank Drake, Jill Tarter, Seth Shostak and Doug Vakoch.

I didn't just want to write a bland congratulatory book. Instead, I decided to take a penetrating look at the aims and assumptions of the entire enterprise. As I wrote it, I kept asking whether we might not be missing something important. Old habits of thought die hard, and a project that has been running for fifty years can benefit from a shake-up. In February 2008 I held a workshop at Arizona State University called 'The Sound of Silence' to encourage radically new ways of addressing the evocative question 'Are we alone?' The contents of this book reflect much of our discussion at the workshop, and my thanks are due to all the participants.

There are some special acknowledgements I should like to make. First and foremost is my wife Pauline Davies, a science journalist and broadcaster with a deeply sceptical mind, and an uncompromising stickler for factual accuracy and properly reasoned argument. She not only pounced on many a slip, but helped me clarify a lot of the arguments, and contributed several ideas of her own which appear without specific attribution in the text. My views on the subject have been greatly shaped by the many in-depth discussions she and I have had over several years. Carol Oliver, former journalist, SETI scientist and now astrobiologist, has been a valued colleague and supporter during my 'SETI career'. Gregory Benford, James Benford, David Brin, Gil Levin and Charles Lineweaver gave good critical feedback on some sections of the book. My literary agent John Brockman has been a decades-long source of encouragement and support for my writing career. My editors Amanda Cook and Will Goodlad have shepherded the project with skill and sympathy; the text is greatly improved as a result of Amanda's detailed critique. And finally, a huge thank you to Frank Drake himself, whose inspirational lectures and articles got me into this field in the first place.

Phoenix, 2009

1

Is Anybody Out There?

Absence of evidence is not the same as evidence of absence.
Donald Rumsfeld (on weapons of mass destruction)

WHAT IF ET CALLS TOMORROW?

On a cold and misty morning in April 1960, a young astronomer named Frank Drake quietly took control of the 26-metre dish at the US National Radio Astronomy Observatory in Green Bank, West Virginia. Few people understood that this moment was a turning point in science. Slowly and methodically Drake steered the giant instrument towards a sun-like star known as Tau Ceti, eleven light years away, tuned in to 1,420 MHz, and settled down to wait.[1] His fervent hope was that alien beings on a planet orbiting Tau Ceti might just be sending radio signals our way, and that his powerful radio dish would detect them.

Drake stared at the pen and ink chart recording the antenna's reception, its fitful spasms accompanied by a hiss from the audio feed. After about half an hour he concluded there was nothing of significance coming from Tau Ceti – just the usual radio static and natural background from space. Taking a deep breath, he carefully reoriented the big dish towards a second star, Epsilon Eridani. Suddenly, a series of dramatic booms emanated from the loudspeaker and the pen recorder began frantically flying back and forth. Drake almost fell off his chair. The antenna had clearly picked up a strong artificial signal. The astronomer was so taken aback he remained rooted to the spot for a long while. Finally, getting his brain in gear, he moved the telescope slightly off target. The signal faded. But when he moved the antenna back, the signal had disappeared! Could this really have been a fleeting broadcast from ET? Drake quickly realized that picking up a signal from an alien civilization on the second attempt was too good to be true. The explanation must lie with a manmade source and, sure enough, the signal turned out to be produced by a secret military radar establishment.

With these humble beginnings – whimsically called Project Ozma after the mythical Land of Oz – Frank Drake pioneered the most ambitious, and potentially the most significant, research project in history. Known as SETI, for Search for Extraterrestrial Intelligence, it seeks to answer one of the biggest of the big questions of existence: *are we alone in the universe?* Most of the SETI programme builds on Drake's original concept of sweeping the skies with radio telescopes for any hint of a message from the stars. It is clearly a high-stakes endeavour. The consequences of success would be truly momentous, having a greater impact on humanity than the discoveries of Copernicus, Darwin and Einstein put together. But it is a needle-in-a-haystack search without any guarantee that a needle is even there. Apart from one or two intriguing incidents (of which, more later) all attempts have so far been greeted with an eerie silence. What does that tell us? That there *are* no aliens? Or that we have been looking for the wrong thing in the wrong place at the wrong time?

SETI astronomers say the silence is no surprise: they simply haven't looked hard enough for long enough. To date, the searches have scrutinized only a few thousand stars within 100 light years or so. Compare this to the scale of our galaxy as a whole – 400 billion stars spread over 100,000 light years of space. And there are *billions* of other galaxies. But the power of the search is expanding all the time, following its own version of Moore's Law for computers, doubling every year or two, driven by surging instrument efficiency and data-processing speed. Now the scope is set to improve dramatically, with the construction of 350 interlinked radio dishes at Hat Creek in Northern California. Named after the benefactor Paul Allen, the Allen Telescope Array will enable researchers to monitor a much larger fraction of the galaxy for alien signals (see Plate 1). The facility is operated by the University of

California, Berkeley, and the SETI Institute, which is where Frank Drake now works. The Institute remains upbeat about the prospects for success, and keeps champagne permanently on ice in anticipation of a definitive detection event.

It's easy to picture the scene if the optimism is right, and something is found soon. An astronomer sits stoically at the controls of the instrument, his feet stuck up on a desk cluttered with papers. Absently, he thumbs though a mathematics textbook. So it has been for him, and dozens of others engaged in SETI, for decade after decade. But today is different. Suddenly the bored astronomer is startled out of his reverie by the shrill, distinctive sound of an alarm. The screech is generated by a computer algorithm designed to spot 'funny' radio signals and separate them from the clutter continually being received from outer space. At first, the astronomer assumes it's just another one of those false alarms, usually a manmade transmission that slips through the net designed to filter out obvious artificial signals coming from mobile phones, radar and satellites. Adhering to the time-honoured protocol, the astronomer keys in some simple instructions and moves the telescope slightly off the target star. The signal immediately dies. He moves the instrument back on target and the signal is still there. After carefully studying the radio wave form and determining that the source remains at a fixed location relative to the stars, the astronomer quickly places a telephone call to a companion observatory involved in the project and simultaneously e-mails the coordinates of the mystery signal.

Five thousand miles away, another astronomer is called out of bed to investigate. Drowsily she wanders to the control room and pours herself a coffee. Then, shaking the sleep from her head, she checks her e-mail and enters the given coordinates. Within a minute the second radio telescope has locked on to the target and immediately picks up the same signal, loud and clear. Her pulse begins to race. Is it conceivable that this time the alert is for real? After decades of unrewarded search, might she be the first person on Earth to confirm that an alien civilization really exists and is transmitting radio signals? She knows that many more checks will be needed before leaping to that conclusion, but the two astronomers, now in excited telephone conversation between different continents, systematically eliminate one mundane possibility after another until, with 90 per cent certainty, they infer that the signal is indeed artificial, non-human and originating far, far out in space. As the radio telescopes continue to track in synchrony and record every minute detail, the dazed pair behave as if in a dream, stunned, awed and euphoric, all at once. What next? Who to tell? What can be gleaned from the data already gathered? *Will the world ever be the same again?*

The story so far (which I admit involves some literary licence[2]) does not demand any great leap of imagination. The basic scenario was well enough portrayed in the Hollywood movie *Contact*, in which Jodie Foster plays the role of the lucky, overawed astronomer. What is far less clear is the next step. What would follow from the successful detection of an alien radio signal? Most scientists agree that such a discovery would be disruptive and transformative in myriad ways. Even contemplating a signal received out of the blue raises many questions: how and by whom would it be evaluated? How would the public get to learn about it? Would there be social unrest, even panic? What would governments do? How would the world's leaders react? Would the news be regarded with fear or wonderment? And in the longer term, what would it mean for our society, our sense of identity, our science, technology and religions? On top of these imponderables is the vexed issue of whether we should respond to the signal, by sending our own message to the aliens. Would that invite dire consequences, such as invasion by a fleet of well-armed starships? Or would it promise deliverance for a possibly stricken species?

There are no agreed answers to any of these questions. The narrative of *Contact* parted company with established science once the signal was received, and lurched off into the speculative realms of wormhole space travel and other dramatic themes. That was science fiction, deriving from the fertile imagination of the late Cornell University astronomer Carl Sagan, author of the book on which the film was based. In the real world, it is completely unclear what would follow the discovery that we are not alone in the universe. In 2001 the International Academy of Astronautics established a committee to address 'what next?' issues. Known as the SETI Post-Detection Taskgroup, its job is to prepare the ground in the event that SETI suddenly succeeds. The rationale is that once a signal from an alien source is confirmed, things would move too fast for the scientific community to deliberate wisely. I happen to be the current Chair of the SETI Post-Detection Taskgroup, and this unusual position has prompted me to give considerable thought to the subject of SETI in general, and post-detection in particular.

IS SETI STUCK IN A RUT?

I've been associated with SETI one way or another for most of my career, and have enormous admiration for the astronomers who operate the radio telescopes and analyse the data, as well as for the technical staff who design and build the equipment. I hope the eerie silence is indeed due to the fact that the search has been limited, and I am a strong supporter of the Allen Telescope Array. But I also think, for reasons I shall come to later, that there is only a very slender hope of receiving a message from the stars at this time, so alongside 'traditional SETI,' of the sort pioneered by Frank Drake, we need to establish a much broader programme of research, a search for *general* signatures of intelligence, wherever they may be imprinted in the physical universe. And that requires the resources of *all* the sciences, not just radio astronomy. There is, however, another factor that has to be addressed. By focusing on a very specific scenario – an alien civilization beaming detectable so-called narrow-band (sharp-frequency) radio messages to Earth – traditional SETI has become stuck in something of a conceptual rut. Fifty years of silence is an excellent cue for us to enlarge our thinking about the subject. Crucially, we must free SETI from the shackles of anthropocentrism, which has hampered it from the very beginning. To help spur this process, I convened a special SETI workshop in February 2008 at Arizona State University's Beyond Center for Fundamental Concepts in Science, with the goal of fostering a lively exchange of ideas between mainstream SETI researchers and a handful of quirky out-of-the-box thinkers, including philosophers, science fiction writers and cosmologists. The upshot was a blueprint for 'new SETI', with some great ideas I shall describe in the coming chapters.

How could something as bold and visionary as SETI become conservative? A major part of the reason is the tendency of humans to extrapolate from their own experience. The very basis for SETI is, after all, an assumption that our civilization is in some respects typical, and that there will be other earths out there with flesh-and-blood sentient beings not too different from us, who will be anxious to communicate. Given that predicate, it is reasonable to take human nature and human society as a model for what an alien civilization will be like – we don't have much else to go on, after all. In the early days of SETI, when the basic strategy was being planned, there were a lot of questions along the lines, 'What would we do in those circumstances?' The result, inevitably, is an inbuilt bias towards anthropocentrism.

Here is a classic example. SETI began with the realization that radio telescopes have the power to beam signals across space. Therefore it's possible that alien signals are coming our way. The image popularized by Carl Sagan was that of an alien civilization directing a message at Earth in the form of narrow-band radio signals. The specifics soon fell into place: the message would be folded into a carrier wave and transmitted from an antenna at a fixed frequency and with enough power to loom above naturally produced radio noise. That is the way terrestrial radio stations do it. It's easy to detect narrow-band signals, once the receiving antenna has been tuned to the right frequency (and, in the case of radio telescopes, pointed in the right direction). There are many other ways to encode and transmit radio messages which require more sophisticated receiving procedures, but SETI astronomers assume that an alien civilization anxious to attract our attention would adopt the simplest method appropriate to entry-level radio technology.

Back in the 1960s, a major preoccupation among SETI researchers was to decide which particular frequency ET might choose, given that there are billions of possibilities. Not all radio frequencies penetrate Earth's atmosphere effectively, so the hope was that the aliens would customize their signals for Earth-like planets by using a frequency that doesn't get greatly attenuated by its passage down from space. But that still left a huge number of potential radio channels. It would be the supreme irony to turn a radio telescope on the right star but tune into the wrong frequency and miss the message. Researchers argued that the aliens would anticipate our dilemma and pick a 'natural' frequency – one likely to be known to all radio astronomers. A popular guess was 1,420 MHz, the emission frequency for cold hydrogen gas. Radio astronomers are very familiar with this pervasive 'song of hydrogen', and it is in some sense a good choice. At any rate, that was the frequency Frank Drake picked for Project Ozma in 1960. Other astronomers suggested multiplying the hydrogen frequency by π, that number being what humans would take to be a 'signature of intelligence' because it enters into both geometry and the equations of fundamental physics, so would surely be familiar to any alien scientist. But there are other special numbers too, like exponential *e* and the square root of 2. In addition, there was a conundrum about whether the aliens would insert a correction to compensate for the motion of their planet and/or our planet.[3] Very soon, the list of possible 'natural' frequencies became depressingly long. However, this battle of the wavebands went away, because technology became available that enables radio astronomers to monitor millions and even billions of radio channels (typically between 1 and 10 Hz wide) simultaneously. As a result, not many SETI researchers worry these days about second-guessing the aliens' choice of frequency. My point is that modest advances in human technology have led within just a few decades to a change in thinking about likely alien communication frequencies. There is a major lesson in this example. It is wise to view the situation through the eyes of the civilization setting out to communicate with us, on the assumption that it has been around for a very long time – at least one million years, and maybe 100 million years or more. Although the aliens may well settle on radio as the medium (perhaps for our benefit), they can hardly be expected to discriminate between 1950s and 1980s levels of human technology: what are a few decades in a million years?

Another case in point: in the 1960s, the laser came to be seen as a powerful alternative means of human communication, and very soon some SETI researchers began to argue that ET, being so much more advanced, would surely prefer to use this fancy tool rather than old-fashioned radio. As a result, optical SETI was born (and still flourishes): astronomers search for a signal in the form of very short-duration, high-intensity pulses of light that with suitable equipment can be distinguished from the overall much brighter but unvarying light of the parent star. Laser communication came less than a century after the invention of radio communication, so once again I ask, what does a century matter to a million-year-old civilization?

A greater degree of parochialism occurs when SETI gets influenced by human politics and even economics. One of the main unknowns is the longevity of a communicating civilization. The challenge is to guess whether ET will be on the air for centuries, millennia or even longer. During the Cold War, many SETI proponents reasoned that the development of advanced radio communication would be paralleled by similar-level technological developments, such as nuclear weapons. Because our own society was at that time in grave danger of nuclear annihilation, it was fashionable to argue that alien technological civilizations likewise wouldn't last long. They would have their own Cold War which, after a few decades, would turn hot, and knock them off the air. When the (terrestrial) Cold War ended, human political concerns shifted to the environment, and SETI thinking duly shifted with it. The hot-button issue now, in many people's eyes, is no longer nuclear war, but sustainability. Transmitting powerful radio waves across the galaxy would require large-scale engineering and soak up a lot of energy. Surely an advanced alien civilization would tailor its technology so as to minimize the environmental impact? Well, maybe. But that line of reasoning would have been received sceptically in the 1960s political atmosphere, and may well be regarded as irrelevant in another hundred years, when environmental problems may be replaced by other concerns. There is no reason to suppose that a million-year-old super-civilization would have 'a sustainability problem'. It might, of course, have some other problem, maybe one we couldn't anticipate, or wouldn't understand even if we were told. SETI is the quintessentially long-term project, and it is foolish to base too much of our search strategy on flavour-of-the-month political fashion. Guessing the political priorities of an alien civilization is futile.

Equally futile is guessing alien economics. Take, for example, H. G. Wells's novel *The War of the Worlds*, in which the Martians, saddled with an inferior planet, consider decamping to Earth. Wells portrays a creepy image of covetous aliens, technologically far ahead of humans, eyeing our planet with malice, '. . . across the gulf of space, minds that are to our minds as ours are to those of the beasts that perish, intellects vast and cool and unsympathetic, regarded this earth with envious eyes, and slowly and surely drew their plans against us.'[4] Wells wrote his story in the 1890s, at the height of the British Empire, when wealth and power were measured in acres of land, tons of coal and iron, and head of cattle. The richest men built railways and owned big ships, mined coal or copper or gold, and purchased vast tracts of grazing land. In short, wealth in Victorian times meant physical stuff. So it was natural to think of alien civilizations similarly valuing real estate and mineral resources, and making plans to spread across space in search of more once their own planet was mined out. Such was the prime motive of Wells's Martians. However, barely a century later, the global economy had transformed out of all recognition. By the 1990s, Bill Gates was the new Rockefeller, making money not from 'physical stuff' but from bits of information. Microsoft had more financial clout than most countries. With information age economics came information age SETI. Surely, it was reasoned, the aliens would not be so primitively rapacious as to scour the galaxy for iron ore, still less for gold or diamonds. An advanced extraterrestrial community would value information – *that* would be their currency, their source of wealth. Information and knowledge – those more noble incentives – would come to dominate the alien agenda. Lust for information may drive them to send out probes, not to acquire material, but to explore and observe and measure, and to compile a database, a veritable *Encyclopedia Galactica*.[5] It seems reasonable enough today, but I wonder how the information argument will play out in the 2090s, when the economy may revolve around something that hasn't yet been imagined, let alone invented. If human priorities can change so dramatically in a mere century, what hope have we of guessing the priorities of a civilization that may have enjoyed a million or more years of economic development?

The same general criticism can be levelled at most theorizing about what an alien civilization would be like and how its members would behave. It's true that the history of human civilization gives a clue, and certain general principles *might* apply to all intelligent life. The problem is, we have only one sample of life, one sample of advanced intelligence, and one sample of high technology. It is really hard to untangle the features that may be special to our planet from any general principles about the emergence of life and intelligence in the universe. In these circumstances there is an inevitable temptation to fall back on analogy with humanity when trying to second-guess ET. But that is almost certainly fallacious. Asking what *we* would do is largely irrelevant. The narrow focus and parochialism inherent in traditional SETI has not been lost on Frank Drake. 'Our signals of today are very different from the signals of 40 years ago, which we then felt were perfect models of what might be radiated from other worlds of any state of advancement,' he writes. 'We were wrong. If technology can change that much in 40 years, how much might it change in thousands or millions of years?'[6] And that's it in a nutshell. However, this clear acknowledgement by the founder of traditional SETI has yet to translate into radical new approaches on the research front. In my opinion, the way forward is to stop viewing alien motives and activities through human eyes. Thinking about SETI requires us to abandon all our presuppositions about the nature of life, mind, civilization, technology and community destiny. In short, it means thinking the unthinkable.

IT'S GREAT – BUT IS IT SCIENCE?

Although the scientific community is on the whole fairly comfortable with SETI these days, members of the public have a hard job positioning it in the scientific landscape. People want to know why it's okay to look for aliens but not for ghosts, why messages from the stars are scientifically respectable, but messages from the dead are not. Where does one draw the line between science and pseudoscience? It is an important but subtle point that goes right to the heart of the scientific method, and it's impossible to understand how SETI works without an explanation of this distinction. So here goes.

Carl Sagan once declared, 'extraordinary claims demand extraordinary evidence.'[7] He made the remark in the context of UFO stories (for which, see the final section of this chapter), but the dictum applies quite generally. Sagan was expressing colloquially what is formally known as Bayes' rule for inference based on the statistical evaluation of evidence. Thomas Bayes was an eighteenth-century English clergyman who appreciated that the weight attributed to evidence will depend on how plausible the hypothesis to which it pertains is deemed beforehand (its so-called prior probability). Let me give an everyday example. I wake at 6 a.m. to find a bottle of milk on my doorstep. What do I conclude? There are two hypotheses. The first is that the milk has been delivered by the milkman, as it is every day except Sunday, because I have a contract with the local company, Express Dairy. Normally the milkman comes at 7 a.m., but perhaps today he came early. The second hypothesis is that the milk has been left there by an altruistic neighbour, Mrs Jones, who might have had a spare bottle. The second hypothesis is obviously a long shot, so it has a much lower prior probability than the first. To believe it, I would require 'extraordinary evidence'. What might that be? Well, Mrs Jones subscribes to the rival company, United Dairies. Their bottles of milk have the brand name 'United' embossed on the side, whereas Express Dairy has 'Express'. If today the bottle displays 'United', I would re-evaluate the odds on the Jones explanation. But I see 'Express'. Do I eliminate Hypothesis 2? Not entirely. It could be that Express Dairy delivered to Mrs Jones by mistake the day before, for example. But the more contrived and extravagant the hypothesis, the greater the weight of evidence needs to be before I will take it seriously. Actually, the probability of either hypothesis being correct is essentially zero, because nobody seems to deliver milk to the doorstep in bottles any more, at least they don't in the countries in which I have lived. So this example is just a bit of nostalgia. (Accurate as of London, circa 1960, for those who are interested. My best friend Brian was the milkman's son, and would occasionally help his father with deliveries. He even recalls turning out on Christmas Day, such was the level of service in the Good Old Days. The milk bottles were originally conveyed to the customer on a horse-drawn cart, and the horse would often get a carrot as a Christmas present. Then the horses were decommissioned in favour of a soulless electrical vehicle. Then the milkman himself was decommissioned, along with the bottles and the vehicle, in favour of horrid supermarket cartons. Such is progress.)

Applied to science and pseudoscience, Bayes' rule helps us assign credibility factors to competing claims. Thomas Jefferson famously said, 'I would sooner believe that two Yankee professors lied, than that stones fell from the sky', when he was told of an eyewitness report of falling meteorites.[8] Like many nineteenth-century intellectuals, Jefferson pooh-poohed meteorite claims on the basis that the deemed prior probability of there being stones in the sky is tiny, whereas the prior probability that a scurrilous professor might make up a story for reasons of fame is not that small. Today we know that the solar system is replete with rubble left over from its formation, so the prior probability we would now assign to a story of a meteorite fall is much greater. We should therefore be inclined to take such reports seriously. (Though still cautiously: a geologist friend of mine has investigated several eyewitness reports of meteorite falls, and they all turned out to be mistaken interpretations.)

A persistent complaint among my non-scientist friends is that modern physics touts all sorts of mind-bending ideas about extra dimensions, unseen dark matter, invisible strings, parallel universes, evaporating black holes, wormholes, etc., in spite of the fact that most of these proposals have little or no experimental or observational evidence to support them. Yet phenomena like telepathy and precognition are experienced first hand by thousands of people, and immediately rejected by scientists as nonsense. Is this not a glaring case of double standards? 'How can you deny the existence of ghosts,' I was once challenged, 'when you accept the existence of neutrinos, which are far more ghostly and have never been seen directly by anybody?' (Neutrinos are elusive subatomic particles that mostly pass right through solid matter, making them exceedingly hard to detect.)

The short riposte to the above complaint is 'Bayes' rule.' The point about modern physics is that weird entities like dark matter or neutrinos are not proposed as isolated speculations, but as part of a large body of detailed theory that predicts them. They are linked to familiar and well-tested physics through a coherent encompassing mathematical scheme. In other words, *they have a place in well-understood theory*. As a result, their prior probability is high. The job of the experimenter is to test the theory. If you build an experiment to make an accurate measurement of such-and-such a quantity, the precise value of which is predicted in advance, then the level of evidence we require to believe that the said entity is real is much less than if someone simply found it by chance in the absence of any theoretical underpinning.[9] Regarding the paranormal, telepathy is not obviously an absurd notion, but it would take a lot of evidence for me to believe in it because there is no properly worked out theory, and certainly no mathematical model to predict how it works or how strong it will be in different circumstances.

So I assign it a very low (but non-zero) prior probability. If someone came up with a plausible mechanism for telepathy backed by a proper mathematical model which linked it to the rest of physics, and if the theory predicted specific results – for example, that the 'telepathic power' would fall off in a well-defined way as the distance increases, and would be twice as strong between same-sex subjects as mixed-sex subjects – I would sit up and take notice. I would then be fairly easily convinced if the experimental evidence confirmed the predictions. Alas, no such theory is on the horizon, and I remain extremely sceptical about telepathy in spite of the many amazing stories I have read.[10]

Turning now to SETI, how does it measure up as science versus pseudoscience? Well, we immediately hit the core problem in the whole enterprise. What prior probability should we assign to the existence of a communicating extraterrestrial civilization? Nobody knows. If you already have good reason to believe ET is out there, and a definite idea about the nature of the signal, then you are, so to speak, 'primed' for the evidence and likely to be easily won over. But if you think the very notion of an alien civilization is incredible, you would need very strong evidence indeed. In Chapter 4 I shall argue that either advanced alien civilizations are very common or they are exceedingly rare: a middle position of a few here and there is intrinsically unlikely.[11] So those who find the notion of alien civilizations a wild and unjustified speculation place SETI in the realm of pseudoscience, while others who find the idea plausible regard it as real science. You, the reader, must make up your own mind. What is not in question, however, is that the *methodology* of SETI is real science. The research is conducted with state-of-the-art technology by highly trained scientists using rigorous techniques of enquiry and analysis, and the results are subject to the usual scrutiny of peer-review. There is no question that the research groups are doing quality science. But are they chasing a chimera? Well, read on . . .

A BRIEF HISTORY OF ALIENS

Speculations about alien beings didn't begin with radio telescopes. Two thousand five hundred years ago, the prophet Ezekiel was walking by the river Chebar in the land of Chaldea when he beheld a glowing whirlwind coming out of the north, from which emerged four weird-looking winged creatures, each superficially 'the likeness of a man'. The creatures were accompanied by four flying wheels that shone like burnished brass, with 'eyes' situated around their rims. Eventually the creatures and the wheels 'lifted up from the Earth' and flew away.[12]

This famous biblical narrative is, of course, just a made-up story, perhaps an account of a dream or vision, perhaps just a colourful way of putting a religious message across. It should not be treated as historical fact, and was presumably never intended as such. Its value lies in revealing to us, through the lens of history, the mind-set of a long-vanished culture. The Israelites, together with many of their contemporaries, firmly believed that mankind was but one form of sentient being in the universe. In most ancient societies, gods, angels, spirits and demons were regarded as real. Many of these non-human beings were thought to be resident somewhere just beyond the sky. All traditional creation myths refer to one or more powerful agents who brought the world into existence, and who continue to visit Earth from time to time.

The idea that humans share the universe with other beings was not just the product of religious mythology; it was also the subject of reasoned argument, as long ago as the fifth century BCE. The Greek philosopher Democritus (460–370 BCE) was an architect of the atomic theory of matter, according to which the universe consists entirely of tiny indestructible particles (atoms) moving in a void. In Democritus' scheme, all forms of matter consist of differing combinations of atoms, and all change is nothing but the rearrangement of atoms. Democritus posited that if nature is uniform, and if atoms can come together in a particular combination to make the Earth, populated by plants and animals, so atoms can arrange themselves in a similar manner in other parts of the cosmos too. Thus he concluded:[13]

> There are innumerable worlds of different sizes. In some there is neither sun nor moon, in others they are larger than in ours and others have more than one. These worlds are at irregular distances, more in one direction and less in another, and some are flourishing, others declining. Here they come into being, there they die, and they are destroyed by collision with one another. Some of the worlds have no animal or vegetable life nor any water.

Democritus' basic argument was vividly captured by the Roman poet Titus Lucretius Carus (99–55 BCE) in his atmospheric *De Rerum Natura*:[14]

> If atom stocks are inexhaustible,
> Greater than power of living things to count,
> If Nature's same creative power were present too
> To throw the atoms into unions – exactly as united now,
> Why then confess you must
> That other worlds exist in other regions of the sky,
> And different tribes of men, kinds of wild beasts.

The birth of scientific astronomy, far from dampening speculations about extraterrestrial beings, actually fuelled them. In the Middle Ages, Copernicus' model of the solar system placed the sun at the centre, and described the planets not merely as wandering points of light, but as other worlds. This transformation encouraged fanciful notions about life on those bodies. In his book *Somnium* (*The Dream*) the astronomer Johannes Kepler went as far as describing a lunar population of reptilian creatures possessing modest intelligence, which he named the Sunvolvans or Privolvans depending on which side of the Moon they dwelt on. He also argued that the Moon 'exists for us on Earth', and therefore the four moons of Jupiter must exist for the Jovians. 'From this line of reasoning,' he declared, 'we deduce with the highest degree of probability that Jupiter is inhabited.'[15] Kepler was not alone in these fanciful notions. The Dutch astronomer Christiaan Huygens produced an entire treatise called *Cosmothereos*, published in its final form in 1698, in which he tried to persuade readers that other planets were inhabited.

Over the subsequent 300 years astronomical observations greatly improved, and the prospects for intelligent life in our solar system dwindled. By the turn of the twentieth century, only one planet remained on the list of candidates – Mars. When I was a high school student, there was a popular belief that the red planet just might be inhabited. It was always the favourite planet for science fiction stories, and the word 'Martian' was almost synonymous with 'alien'. Mars is certainly not a write-off as an abode for life. Admittedly it is smaller than Earth, so has a lower gravity, and is situated farther from the sun, making it cold. On the other hand, it does possess an atmosphere, albeit thin, and the surface temperature can sometimes rise above the freezing point of water. By the middle of the nineteenth century, telescopes were large enough to reveal many surface features. Astronomers saw polar caps grow and shrink, and seasonal changes in colour that hinted at vegetation.

In 1858, Angelo Secchi, a Jesuit monk in Italy, began mapping Mars and named some of the vaguely linear features *canali*, meaning channels. Twenty years later his compatriot, the astronomer Giovanni Schiaparelli, produced improved maps of Mars, and also used Secchi's term *canali*. The sobriquet became liberally translated into English as 'canals', with the hint of artificiality. The 'canals' of Mars caught the imagination of a wealthy American writer and traveller, Percival Lowell, who built an observatory at Flagstaff in Arizona dedicated to studying Mars and seeking evidence for life. By 1900, Lowell was convinced he could discern signs, not just of life, but intelligent life. He started making elaborate drawings displaying complex networks of lines, which he took to be aqueducts built by an advanced civilization to convey melt-water from the polar caps to the parched equatorial regions (see Plate 2). At about the same time, H. G. Wells wrote his masterpiece, *The War of the Worlds*.

At the time Wells and Lowell published their works, it was not unreasonable to believe that Mars could host intelligent life, a notion that lingered in some quarters right up to the dawn of the space age. Then, in 1963, NASA sent a space probe called Mariner on a Mars fly-by. The pictures that came back showed a barren, heavily cratered landscape, more resembling the Moon than Earth. Follow-up Mariner probes measured a disappointingly low atmospheric pressure and found no trace of oxygen. Without oxygen there can be no ozone layer, so the surface of Mars is subjected to withering ultraviolet radiation from the sun. Bitter cold, a tenuous atmosphere and a surface awash with ultraviolet radiation add up to a pretty lethal combination, so hopes for life on Mars began to fade. Significantly, Mariner found no trace of the famous canals, although it did photograph dried-up river systems. Lowell's canals turned out to be a figment of his fertile imagination, a case of wishful thinking rather than scientific data. It is a salutary lesson that is well worth remembering when considering the subject of SETI.

LIFE AMONG THE STARS

Today, we can be pretty sure that there are zero prospects for intelligent life arising on any other planet in the solar system. SETI, however, targets extra-solar planets. When Drake started Project Ozma, this represented something of a leap of faith, because astronomers at that time couldn't be sure there *were* any planets beyond the solar system. It has only been in the recent past that some have been identified. To date, about 400 have been found orbiting stars in our immediate neighbourhood of the galaxy. Two methods have produced the majority of discoveries. The first depends on the fact that a planet exerts a force on its parent star, making the star wobble very slightly in its motion. Careful study of the star's light will detect this movement as a periodic shift in wavelength (known as the Doppler effect). Another technique looks for slight changes in the brightness of a star caused when a planet crosses its face (known as the transit method). At this time, only one extra-solar planet has been photographed as an object recognizably distinct from the parent star. The reason it is so hard to capture an image is that the glare of the star totally swamps the feeble light from the planet; it's like trying to detect a firefly next to a searchlight. Because both the Doppler and transit methods work best for very massive objects orbiting close to the star (dubbed 'hot Jupiters' by the popular press), few of the planets so far identified this way are Earth-like. Recently, several 'super-earths' have been catalogued; these are relatively small dense planets, but with masses still several times that of Earth's. Nevertheless, astronomers mostly agree that there should be abundant earth-sized planets out there, and they look forward to better optical systems that will one day image these 'other earths' in detail. Meanwhile, a satellite called Kepler, launched in March 2009, is monitoring 100,000 stars continuously over three years for transits. Kepler has the sensitivity to detect, although not to photograph, planets small enough to resemble Earth.

From the standpoint of hosting life, it's not sufficient that a planet has roughly the same radius as Earth. To be truly 'Earth-like' involves several other features thought to be essential to biology. For example, the planet must possess a reasonably thick atmosphere. It probably also needs a hot interior, both to generate a magnetic field for deflecting hazardous cosmic radiation, and to drive plate tectonics (movement of continental crust), which is important for recycling chemicals in the surface environment. Undoubtedly the most crucial requirement for life as we know it is liquid water: no known life can function without it. These conditions have led to the concept of 'the habitable zone' – a region of space around a star where the surface of a planet could support liquid water. In the case of the solar system, the habitable zone extends from somewhere between Venus and Earth (Venus is far too hot for liquid water), out to about Mars (which is mostly, but not always, too cold).

To be 'in the zone' ideally requires an Earth-like planet in an Earth-like orbit around a sun-like star. However, the traditional view of habitable zones is now recognized as overly restrictive and needs to be enlarged to include some interesting additional possibilities. For example, a cool star such as a red dwarf could possess a narrow small-radius habitable zone. In 2007 a planet that might support life was discovered around a red dwarf named Gleise 581. The planet is a super-Earth, orbiting a mere 11 million kilometres (7 million miles) from the star (compare Earth, 150 million kilometres (93 million miles) from the sun). That is close enough for water to be liquid even though the star is dim. Unfortunately for advanced life, a planet that close to a star is certain to be phase-locked – with one side stuck facing the star, much as the Moon is phase-locked to Earth (we can't see the far side of the Moon from Earth). Phase-locking implies that half the planet is permanently sweltering and the other half permanently frozen, which is not an ideal arrangement for biology. There will, however, be a Goldilocks zone at the margins where primitive life at least might be possible.

Yet another variety of habitable zone would be the interior of small icy planets or moons. In the frigid outer suburbs of our own solar system, Europa, a moon of Jupiter, has a liquid ocean beneath its ice crust, warmed by tidal friction from Jupiter's gravity (see Plate 3). Farther out, the dwarf planet Pluto is now known to be but one member of a large class of icy bodies, some of which are also rich in life-encouraging chemicals. The larger ones have enough interior heat from their formation, plus the warming effects of radioactivity and chemical processes, to remain liquid inside for billions of years. Other planetary systems will almost certainly contain similar bodies with frozen surfaces and liquid-water interiors. If life were to emerge inside these ice-capped bodies, it would most likely be stuck at the level of microbes. But even if more complex biological entities were to evolve there, one can only speculate what life would be like in such a location. How long would it take sentient beings, confined to their pitch-dark liquid habitat by a solid sky hundreds of kilometres thick, to discover that there was a vast universe beyond their world's apparently impenetrable roof? It is hard to imagine that they would ever 'break out' of their ice prison and beam radio messages across space.

AND FINALLY, WHAT ABOUT ALL THOSE UFO STORIES?

Surveys show that a staggering 40 million Americans have seen what they describe as a UFO. So what is a UFO? The acronym means Unidentified Flying Object, so it literally means nobody knows what it is. But the press has turned a negative – we don't know – into a positive – we know it is . . . Something Else. In the popular imagination, that something else is a spaceship from another world. So if someone sees something in the sky they can't identify, then – so the popular argument goes – it is a candidate for an alien spacecraft.

Needless to say, none of this impresses scientists. For a start, the logic is flawed. Not being able to identify something as X, doesn't mean it must be Y. It might be Z. UFOs are reported in their thousands, and the vast majority of them get explained straightforwardly as weird atmospheric effects, aircraft seen under unusual conditions, bright planets, etc. Admittedly, there are a handful of tough cases, but no obvious demarcation divides cases that get solved from those that don't. So it is tempting to conclude that if 95 per cent of sightings can be explained without too much effort, then so could the remaining 5 per cent if we had enough information at our disposal, because there is nothing to elevate that residue from the rest, apart from being more puzzling.

This is certainly the position of many governments that have set up UFO research studies. The British government recorded 11,000 cases starting in 1950. After downplaying the importance of this study for years, it recently released a large batch of UFO files under the Freedom of Information Act. But in spite of some baffling cases, the government's conclusion was that, whatever the unexplained residue might represent, it was not aliens at work. 'The Ministry of Defence does not deny that there are strange things to see in the sky,' conceded a spokesperson. But on the other hand . . . 'It certainly has no evidence that alien spacecraft have landed on this planet.'[16]

For its part, the United States established Project Blue Book in 1950 to evaluate whether UFOs posed a threat to national security. Over twenty years, thousands of reports were sifted and hundreds investigated in detail. At the end of this mammoth analysis, Edward Condon, a well-known atomic physicist, was asked to provide an assessment. The resulting Condon Report concluded that about 90 per cent of the sightings could be explained in terms of normal phenomena, while the remaining 10 per cent didn't contain enough of scientific value or defence significance to warrant Blue Book's continuation.[17] It was duly terminated. Blue Book employed an astronomer as scientific adviser – Allen Hynek from Northwestern University in Illinois. I met the amiable pipe-smoking Dr Hynek on a number of occasions when I was a postdoctoral researcher, and I even visited his home in Illinois, which contained a room full of dusty UFO files. That was in 1970. It was Hynek who sorted the reports into various categories and coined the familiar term 'close encounters of the third kind', which became a byword after Steven Spielberg adopted it for his famous movie (and in return gave Hynek, complete with pipe, a cameo role in the film). Hynek was convinced after years of gruelling investigation that there was 'something in it', although he conceded that only a tiny fraction of cases presented evidence for anything seriously odd. For a while he almost convinced me too – I was at least prepared to keep an open mind. But over the years, as I thought more about these unexplained sightings, I came to see how deeply anthropocentric they were – bearing all the hallmarks of human rather than alien minds. This was especially true of the most challenging cases in which witnesses claimed to have encountered alien beings in the flesh. Almost always these 'ufonauts' were humanoid in form (sometimes dwarfs or giants), and often with descriptions that suggested something straight out of

Hollywood central casting. Later I shall discuss how plausible this is – that alien spacefarers would resemble humans so closely in their physical form. Another giveaway was the banality of the aliens' putative agenda, which seemed to consist of grubbing around in fields and meadows, chasing cows or aircraft or cars like bored teenagers, and abducting humans for Nazi-style experiments. Not what one would expect of cosmic superminds.

From time to time I have solved a few cases myself. Some were easy. One consisted of a movie showing a bright light rising from the ground in the east just before sunrise, and gradually fading from view in about half an hour. As any amateur astronomer would immediately know, this was Venus, presenting itself as the 'Morning Star', rising ahead of the sun right on cue. Another movie showed a set of lights against a cloudy sky, each one lazily falling with a slight rocking motion before blinking out. The film had been taken by a couple camping near Stonehenge in southern England, a location redolent with ancient folklore and mystical ambience. If you are going to see UFOs, there is no better place. The film looked so striking that Granada Television showed it on the national 6 p.m. news, and organized a live interview to follow. I was asked to take part. I reached the studio early and naturally asked for a sneak preview. The moment I saw the movie sequence I knew what the lights were – military flares. This was pure luck on my part: I had witnessed something very similar myself not long before. I asked the studio operator to zoom in on the images and, sure enough, there were the smoke trails. The flares had been ignited above the cloud base, and then emerged on little parachutes, swaying in the wind, so that they appeared one by one from the clouds and slowly descended before eventually burning out. Once the explanation was presented, the lights no longer looked so mysterious. The fact that Stonehenge is located close to a British army training ground hadn't occurred to anybody as significant. Granada TV unsuccessfully tried to pull the story once the explanation was clear. The live show went ahead too, so I asked the witnesses to describe the scene. Apparently they had observed the strange lights in the same patch of sky for several days running before filming them. I wanted to know why they didn't get in closer if the phenomenon was so predictable. 'We tried,' they replied, 'but were prevented by the army, who were conducting manoeuvres in the area.'

Now you might think that, given all this, my military flares explanation would immediately have won the day, but not a bit of it. In the eyes of the couple, and probably the majority of the viewing public too, the objects in the film really were UFOs, it's just that they *looked like* military flares. With that sort of reasoning, you can't win.

Of course, the same is true of all conspiracy theories. Many people are convinced that 'the government' knows 'the truth' about UFOs but is covering it up for nefarious reasons. This is superficially plausible, because governments certainly do have a habit of covering things up. I asked Seth Shostak of the SETI Institute in California, who has studied the UFO scene in detail, what he thought about it. 'Would they really be so efficient at covering up a big thing like this?' he replied sceptically. 'Remember, this is the same government that runs the Post Office.' He also pointed out that UFOs are not the exclusive preserve of the United States: they are reported worldwide. It's not enough for the US government to conceal the truth over many decades. What about the governments of, say, Belgium or Botswana? You might expect at least one of them to let something slip from time to time.

None of this constitutes a knock-down 'solution' of the UFO 'riddle'. It would not surprise me if a small fraction of cases involve new or little-understood atmospheric or psychological phenomena. But whatever lies behind that stubborn residue of hard-to-explain cases, I see no reason to attribute them to the activities of alien beings visiting our planet in flying saucers. UFO stories, like ghost stories, are fun to read, but cannot be taken seriously as evidence for extraterrestrial beings. They do serve a useful purpose, however, by providing a window on how the human mind imagines aliens and alien technology. What is striking about the accounts is not their weird and otherworldly character, but their distinctly mundane and human-like quality. We would surely expect of extraterrestrials something more extraordinary than humanoid beings piloting the equivalent of souped-up stealth bombers.

As I shall show, SETI compels us to make *much* greater leaps of imagination. The British biologist J. B. S. Haldane famously remarked that 'the universe is not only queerer than we suppose, but queerer than we *can* suppose.'[18] Contemplating a seriously alien intelligence, and the hallmarks of a multi-million-year technology, means we must jettison as much mental baggage as possible. Forget little green men, grey dwarfs, flying saucers with portholes, crop circles, glowing balls and scary nocturnal abductions. Embracing SETI means going beyond UFOs, beyond the stereotypes of human myth, beyond folklore, fable and science fiction. Even Oz, the fantasy land after which Drake named Project Ozma, is not 'queer enough', to paraphrase Haldane. To fully comprehend the significance of the eerie silence compels us to embark on a journey into the *truly* unknown.

2

Life: Freak Side-Show or Cosmic Imperative?

We now know the number of stars in the universe is something like one followed by 23 zeros. Given that number, how arrogant to think ours is the only sun with a planet that supports life, and that it's the only solar system with intelligent life.

Edward J. Weiler, NASA Director[1]

A UNIVERSE TEEMING WITH LIFE?

Most people have little difficulty accepting that there may be countless inhabited worlds scattered through space. When asked to justify this belief, a typical response is that the universe is so vast, there simply *must* be life and intelligence out there somewhere. It is an oft-repeated argument, but unfortunately it contains the elementary logical fallacy of confusing a necessary with a sufficient condition. Consider the two basic requirements for life to exist on an Earth-like planet: first, the Earth-like planet; second, the genesis of life. Suppose we grant that there are indeed trillions of Earth-like planets in the observable universe – a prospect that is looking increasingly likely – does this guarantee trillions of inhabited planets? Not at all. The fact that a planet is *habitable* is not the same as saying it is *inhabited*. That would be so only if the genesis of life is guaranteed, given that a planet is Earth-like. But suppose the emergence of life from non-life is a freak affair, an event of such low probability that even with a *trillion trillion* habitable planets it would still be unlikely to happen more than once? The sheer size of the universe would then count for little if the odds are so heavily stacked against the spontaneous formation of life.

What do we know about life's origin? Might it have been a bizarre fluke, a one-off accident making Earth unique in the observable universe? Many distinguished scientists have thought so. Francis Crick, co-discoverer of the structure of DNA, once wrote, 'The origin of life appears at the moment to be almost a miracle, so many are the conditions which would have had to have been satisfied to get it going.'[2] Jacques Monod, the French biochemist who won a Nobel Prize for his work unravelling the details of the genetic code, similarly proclaimed, 'The universe is not pregnant with life nor the biosphere with man . . . Man at last knows that he is alone in the unfeeling immensity of the universe, out of which he emerged only by chance.'[3] At that time, belief in any form of extraterrestrial life, let alone intelligent alien beings, was seen as pure science fiction, the stuff of bad Hollywood movies, with no scientific basis whatsoever. I was a student in the 1960s and my own fascination with the possibility of extraterrestrial life was regarded as so disreputable it verged on the crackpot. One might as well have expressed a belief in fairies. SETI in particular wasn't taken seriously. The distinguished Harvard biologist George Simpson described the search for intelligent aliens as 'a gamble at the most adverse odds with history'.[4]

Today the pendulum has swung the other way. The biologist Christian de Duve – like Monod, a Nobel prizewinner – is so convinced that life will arise on Earth-like planets throughout the universe, he calls it 'a cosmic imperative'.[5] Both scientists and journalists now often declare that the universe is chock-a-block with life. Every little discovery concerning planets is presented by the media as one step closer to finding extraterrestrial life, even intelligent life. The 2009 meeting of the American Association for the Advancement of Science, held in a snow-covered Chicago just before the launch of the Kepler mission to search for Earth-like extra-solar planets, typified the new mood. Several sessions were devoted to astrobiology – a subject that includes the study of life beyond Earth. In one of them, Alan Boss of the Carnegie Institution in Washington, DC, declared in ebullient fashion: 'If you have a habitable world and let it evolve for a few billion years then inevitably some sort of life will form on it . . . It would be impossible to stop life growing on these habitable planets.' Boss went on to deliver an arresting statistic: 'There could be one hundred billion trillion

Earth-like planets in space, making it inevitable that extraterrestrial life exists.'[6] The science journalist Richard Alleyne reported this event for the UK's *Daily Telegraph* newspaper: 'Life on Earth used to be thought of as a freak accident that only happened once. But scientists are now coming to the conclusion that the universe is teeming with living organisms.'

So which point of view is right? Is life a freak accident, confined to our planet, or a 'cosmic imperative', and hence spread throughout the universe? The answer hinges on just how likely it is for life to emerge from non-life, so it makes sense to look for clues in the way that life on Earth began.

HOW DID LIFE BEGIN?

When Charles Darwin published his magnum opus *On the Origin of Species*, he gave a convincing account of how, over immense periods of time, life has evolved from simple microbes to the richness and complexity of the biosphere we see today. But he pointedly left out an account of how life got going in the first place. 'One might as well speculate about the origin of matter,' he quipped. Two centuries later we are still largely in the dark about how life started.

There are really three puzzles rolled into one here – the when, where and how of biogenesis. The when part at least is becoming clearer. After some academic skirmishes over the past decade, most biologists agree that the Pilbara hills of Western Australia contain traces of life dating back nearly 3.5 billion years.[7] Now a focus of intense international research, the ancient rocks jut from arid hillsides in a wild and desolate terrain about four hours' drive through the bush from the coastal town of Port Headland. The evidence for life gathered so far includes fossilized microbial mats called stromatolites and tiny features embedded in rock, thought by many researchers to be microfossils. Recently, evidence has been found in the same region for an entire fossilized ecosystem.[8]

Could life have existed at an even earlier epoch? The problem in answering this question is the paucity of very old rocks. There are some in Greenland that have been dated to 3.85 billion years ago, which are subtly altered in a manner consistent with biological activity, but non-biological processes could also be responsible. Rocks even older than this are known, but so far none has been found to contain any trace of ancient life. Obviously the Pilbara organisms didn't just pop into existence ready-made; there would have been a period of evolution preceding their appearance. All we can say with confidence is that life had established itself on Earth by some time between 3.5 and 4 billion years ago. This may be compared with the age of the planet itself – about 4.5 billion years.

As to where life began, that is much more problematic. The Pilbara hills provide the earliest clear traces of life on Earth, but there is no reason to suppose life actually started there. Darwin himself mused about a 'warm little pond' full of chemicals leached from the surrounding rocks and energized by sunlight. Various other types of 'primordial soup' have been suggested, ranging from drying lagoons through suspended water droplets to the entire ocean. Other researchers favour the vicinity of the scalding fluids spewing from deep-ocean volcanic vents. My own favourite locale, for what it's worth, is far beneath the seabed (maybe as deep as a kilometre or two) in the pores of rocks infused by slow currents of hot convecting fluid. In truth, the setting is pure guesswork. It is not clear that life even began on Earth; a good case can be made that it started on Mars, for example. Earth and Mars have for billions of years traded rocks blasted into space by comet and asteroid bombardment, and the surface of Mars is pockmarked with impact craters. Much of the ejected material goes into orbit around the sun, and a small fraction of that eventually hits Earth, perhaps after a million years or more in space. Over the course of geological history, trillions of tons of Martian material have rained down on our planet. It is but a small step to imagine Martian microbes hitching a ride on some of this debris.[9] Embedded deep within a rock, protected from the harsh conditions of space, a hardy microbe could easily survive the interplanetary journey, especially if it was in a spore-like dormant state. Experiments have confirmed that microbes inside rocks can withstand space conditions, as well as blast-off and subsequent high-speed entry into Earth's atmosphere.[10]

Why Mars? The case for life starting there first is not overwhelming, but it is at least suggestive. Mars is a smaller planet, so it cooled quicker from the heat of formation, and hence was ready for life sooner than Earth. For about 700 million years both planets were ferociously pounded by objects ranging in size from small boulders to massive 500-kilometre-wide asteroids. The surface layers churned up by the bombardment are more loosely packed on Mars than on Earth owing to the lower gravity, and so would have offered a deeper refuge from the mayhem for any subsurface microbes. Mars does have water, but not much. Its relative scarcity might actually have been a help for life's early survival: on Earth, the heat energy released by the biggest impacts boiled the oceans and swathed the planet in a lethal atmosphere of rock vapour and superheated steam. Today, Mars is a freeze-dried desert, at best only marginally habitable to terrestrial microbes, but billions of years ago the tables were turned: Mars was more favourable to life, with streams and lakes, a much thicker atmosphere and higher surface temperatures than today. None of this adds up to a convincing case that life on Earth came from Mars, but it does widen the range of settings that need to be explored in answering the question of where life began.

The really tough problem about the origin of life is *how* it happened. It is easy to appreciate the basic obstacle. The simplest known life form is already so immensely complex it is inconceivable that such a thing could have arisen spontaneously in a single transformation purely by chance. In a famous metaphor once used by the British astronomer Fred Hoyle, it is easier to believe that a whirlwind passing through a junkyard would assemble a functioning Boeing 747.[11] However, the operative word here is 'known' life. Nobody supposes the first living thing was as complex as a bacterium. Far simpler forms of life may be possible, providing stepping stones from the first organism to life as we understand it today. It could be that these primitive bugs are still out there somewhere, overlooked for what they are, either too small to have attracted attention or confined to a peculiar habitat that hasn't yet been explored by microbiologists (of which, more later). They may even have been left behind on Mars. It is equally conceivable that simpler precursors of familiar life long ago died out, either gobbled up or elbowed aside by more complex, sophisticated life, leaving no trace.

Life (at least as we know it) is chemical in nature. That may seem

Fig. 1. Life in a test tube? Stanley Miller and his famous organic synthesis experiment.

obvious, but in the subject of SETI nothing should be taken for granted. Two hundred years ago life was regarded as some sort of magic matter, animated by a mysterious vital force. Scientists still use the term 'organic chemistry', even though we now know that the laws of chemistry are the same whether a molecule is located inside or outside an organism. Most of the early speculation about the origin of life, such as Darwin's warm little pond, assumed there was a well-defined chemical pathway – perhaps long and tortuous – between an amorphous chemical cocktail and the first organized living cell. Life's origin would then be akin to baking a cake: there would be a recipe of required substances, and a procedure – heating, drying, cooling, etc. – for transforming non-living stuff into life. It is a beguiling concept, and one reinforced by a famous experiment conducted in 1952 by Stanley Miller at the University of Chicago. At the instigation of the geochemist Harold Urey, Miller filled a flask with methane, water, ammonia and hydrogen – gases thought at the time to have been present in Earth's primitive atmosphere – and sparked the mixture with electricity for a few days. Miller was delighted to discover amino acids, the building blocks of proteins, in the sludge at the bottom of the flask (see Fig. 1).

The Miller–Urey experiment came to be seen by many chemists as the first step on the long road to synthesizing life in the laboratory, re-creating the same chemical pathway that Mother Nature took billions of years ago. Unfortunately that entire line of research, which looked so promising in the 1950s, turned out to be something of a dead end. Amino acids are undeniably building blocks of proteins, but they are as far from the completed product as a brick is to the Empire State Building. Also, they are easy to make, and are found occurring naturally in meteorites and even in interstellar dust clouds. Going beyond amino acids, let alone producing nucleic acids (the basis of heredity), has proved impossible using a simple energized soup procedure. If life was incubated by successive chemical transformations, it was unlikely to be in this straightforward manner.

Since Miller–Urey, our understanding of the nature of life has undergone a revolution. In that same year, Francis Crick and James Watson published their paper on the structure of DNA, and in subsequent decades scientists have come to regard the living cell less as magic matter, more as supercomputer. To be sure, life uses chemistry to enact its agenda, but the key to its near-magical qualities lies with the way cells process and replicate information. That puts a different complexion on the whole biogenesis puzzle, because the real issue is how information storage and replication might have arisen spontaneously, not how naturally occurring chemicals reacted to 'animate' matter.

Obviously a crucial part of this story is complexity. To qualify for the description 'alive', a system has to do more than merely replicate information (a simple salt crystal can do this): it needs to be complex enough to possess a type of autonomy. That is, the information content has to be great enough for the system to manage its own agenda – to 'take on a life of its own', quite literally. It is far from clear what that threshold of complexity might be, but the simplest known naturally occurring autonomous microbes each contain upwards of a million bits of information. Areas of research that have a bearing on the problem are the study of self-organizing systems, the self-assembly of molecular structures, complexity and information theory in general, and a burgeoning field of investigation known as synthetic biology, in which researchers endeavour to design and make their own organisms from scratch in the laboratory. These are exciting and fast-moving fields, but all that can be said at this time is that the problem of life's origin is very far from being clearly formulated, and nowhere near being solved.

Even if we never know exactly how life began, however, we might still solve the lesser riddle of whether its origin was a fluke or a likely event. From the point of view of SETI, all we really need to know is whether life starts up readily and is therefore widespread in the universe, as seems to be so widely believed.

LIFE AS A BIZARRE FLUKE

To a physicist like me, life looks to be little short of magic: all those dumb molecules conspiring to achieve such clever things! How do they do it? There is no orchestrator, no choreographer directing the performance, no esprit de corps, no collective will, no life force – just mindless atoms pushing and pulling on each other, kicked about by random thermal fluctuations. Yet the end product is an exquisite and highly distinctive form of order. Even chemists, who are familiar with the amazing transformative powers of molecules, find it breathtaking. George Whitesides, Professor of Chemistry at Harvard University, writes, 'How remarkable is life? The answer is: *very*. Those of us who deal in networks of chemical reactions know of nothing like it.'[12] Whitesides stresses how hard it is to imagine such a complex and specifically organized system coming into being in the first place: 'How could a chemical sludge become a rose, even with billions of years to try?[13] . . . We (or at least I) do not understand. It is not impossible, but it seems very, very improbable.'[14] Which brings us to the crux of the matter: just *how* improbable is it? The entire SETI enterprise hinges on the answer. Whitesides again: 'But how likely is it that a newly formed planet, with surface conditions that support liquid water, will give rise to life? We have, at this time, no clue, and no convincing way of estimating. From what we know, the answer falls somewhere between "impossibly unlikely" and "absolutely inevitable". We cannot calculate the odds of the spontaneous emergence of cellular life on a plausible prebiotic earth in any satisfying and convincing way.'[15]

It might have been different had the arrangement of chemicals in the cell followed some sort of pattern; for example, if the sequences of amino acids that make proteins contained mathematical regularities that could be traced back to an underlying law of nature. But no such orderliness is apparent: the chemical sequences seem totally haphazard, which was what led Monod to his bleak conclusion. Yet they are not arbitrary: in many cases even a small change in the sequence can severely compromise biological functionality. So the arrangement is at once *both* random *and* highly specific – a peculiar, indeed unique, combination of qualities hard to explain by deterministic physical forces.[16] On the other hand, if chance dominates when it comes to the origin of life, the odds in favour of getting just *that* arrangement of molecules are infinitesimal – the tornado in the junkyard. Viewed this way, then, life is a freak phenomenon that arose by an exceedingly lucky fluke, a process of such staggering improbability that we can safely say it happened only once in the observable universe. The fact that we are witness to such a near-miracle is, of course, not at all a surprise, but an inevitable selection effect: observers can exist only where there is life.[17]

In spite of these dampening facts, belief in extraterrestrial life is now widespread among scientists. So what has changed since the days of pessimists like Crick, Monod and Simpson? Curiously, very little on the actual scientific front. It's true that we can now be reasonably sure there are lots of planets in the universe, but that merely confirms what astronomers already suspected in the sceptical sixties. Since then, some basic organic molecules have been found in space – in comets and molecular clouds – but as I have explained, making the building blocks of life is easy, and has very little relevance to the problem of how to assemble them into highly complex arrangements characteristic of life, let alone in a manner that systematically processes information. Perhaps the most pertinent change is the discovery that micro-organisms can withstand a wider range of conditions than was obvious a few decades ago, implying that more planets could in principle support simple life. But this only increases slightly the range of planets we might regard as qualifying for the accolade 'Earth-like'. It doesn't alter a jot the fact that life's origin could have been a freak event.

Much ado is made about finding signs of liquid water – on Mars, for example. NASA has an unofficial mantra, 'follow the water', as if life will be obligingly waiting wherever we find a lake or an ocean. It is often pointed out that where there is liquid water on Earth, there is life. It's true that liquid water is essential for life as we know it, but the sequence of reasoning planets→water→life is another glaring example of confusing a necessary with a sufficient condition. Liquid water may indeed be necessary for life, but it is far from sufficient: there may be a host of other conditions that are also required. On Earth, we find life in almost all liquid water habitats not because it has arisen spontaneously there, but because Earth's hydrosphere forms a more or less contiguous system, so life has been able to spread out and invade all those watery places. Following the water into space isn't misconceived, but it is similar to the man who loses his keys in the dark and looks for them under the lamppost, not because they are likely to be lying there, but because there is no chance at all of finding them anywhere else.

None of the scientific discoveries of the past half-century have greatly altered what we know, or don't know, about life's seemingly freaky nature. The change in sentiment is due, I believe, to fashion rather than discovery. At a time when physicists freely speculate about extra dimensions, antigravity and dark matter, and cosmologists propose multiple universes and dark energy, speculation about extraterrestrial life seems tame by comparison. I'm okay with that. It's fun to speculate, and ET may indeed be out there somewhere. Or not. However, we must never allow speculation to replace real science.

One way to bring real science to bear on this subject is to see whether de Duve's 'cosmic imperative' stacks up. Could it be that the laws of nature are in some way rigged in favour of life, making its emergence far more likely than the mere random shuffling of molecules might imply? The answer is no, at least not at first glance. I already mentioned that there is no discernible pattern in the sequences of amino acids in proteins. The same goes for the sequences of base-pairs – the 'genetic letters' – in DNA. It all looks random. If the laws of physics and chemistry are somehow conspiring to fast-track matter to life against the raw odds, it's not showing up in the end product – the molecular structures themselves. Indeed, the laws of physics and chemistry are completely indifferent to the sequences of base-pairs in DNA or amino acids in proteins: they display no favouritism for one sequence over another.[18] Commentators often declare that life is 'written into' the laws of nature, but if it is written into the laws of physics and chemistry we have yet to see any sign of it. This comes as no surprise to a physicist. The laws of physics are, after all, universal. They are no more likely to have 'life' written into them than 'laptop computers' or 'the Rocky Mountains'. Life, computers and mountains are consistent with the laws of physics, but the laws alone do not explain their existence.

Does this invalidate the cosmic imperative? Not necessarily. The basic laws of physics may not exhaust all possible laws. For example, there are law-like regularities of a quite general nature describing complex self-organizing systems as diverse as ant colonies, stock markets and the internet. These 'organizational' laws *augment* those of fundamental physics; they don't supplant or override them. It could be that life is the product of such a higher-level (or emergent) law, perhaps a law of increasing complexity that operates, not universally like the laws of physics, but in special (though not especially improbable) systems satisfying as yet unknown conditions. If so, then all it might need is for chance to create such a special system in the first place following which the law would serve to drive it towards life. Personally I have long been attracted to the possibility of such higher-level laws, e.g. laws of increasing complexity, but I freely admit that there is scant evidence for them so far.[19] I shall return to this topic in Chapter 8.

Another line of reasoning in favour of the cosmic imperative comes from a variety of mathematical games in which 'lifelike' behaviour seems to emerge quite effortlessly even when the rules of the game are very simple. One class of games, called cellular automata, offers a cartoon world in which squares on a chequerboard are filled or not so as to form a pattern; the pattern then evolves deterministically according to simple rules. A particular cellular automaton, devised by the British mathematician John Conway in 1970 and known appropriately enough as *The Game of Life*, has become quite fashionable, and exhibits a remarkably rich and complex ecology of shapes that move and interact.[20] If simple processes 'played' in combination can generate accelerating organized complexity, maybe the secret of life isn't so subtle after all. On the other hand, real life seems as far from *The Game of Life* as a mouse is from Mickey Mouse. Simple mathematical representations are great fun, but they mustn't be confused with reality. At best, cellular automata tip the scales slightly in favour of the idea that life starts up easily.

Although nothing like a 'life principle' has been identified buried in the laws of physics and chemistry, biologists agree that there is at least one organizing principle under-girding all of life: Darwinian evolution. Any system that undergoes replication with variation and is subjected to natural selection will evolve over time. This principle, which is really a truism (it merely states that entities which replicate more efficiently increase their relative numbers in the population), can be taken as a definition of life. Evolution can, but does not have to, lead to greater complexity. So life *may* have begun with something comparatively simple – a population of small replicating molecules, say. Perhaps these molecules are simple enough to form spontaneously in many environments; they may even be forming on Earth today. Once the initial molecular replicators get going then Darwinian evolution can kick in, driving the complexity higher and higher, until something approaching the familiar living cell eventually emerges. The important point is that Darwinism doesn't have to wait for cellular life to arise before it can work its spell; it could be equally effective at the molecular level. This claim is easy to make, but it leaves a lot of questions open, not least of which is the identity of the first replicators. What are these molecules, exactly? Nobody knows, although the chemist Graham Cairns-Smith has conjectured they may not even be organic molecules; he favours impure clay crystals.[21]

Actually, it's not strictly necessary for life to begin with replicating structures at all. All that is required is the replication of *information*. Bits of information can be represented whenever there is a pattern in a physical structure. The pattern can be replicated either by reproducing the structure itself, or by merely copying the pattern on to a 'blank'. For example, when I transfer a computer file from a memory stick on to an empty part of the hard drive of my computer, the computer doesn't make a physical copy of the inside of the memory stick. What happens is that the bits of information (i.e. the electrical pattern) in the stick get copied on to the hard drive. It is the *software* that is replicated, not the hardware. Life could begin simply by patterns being copied, with small variations, and subjected to selection pressure. The patterns could be anything at all, e.g. complex magnetic or electrical tessellations or arrays of spinning atoms, coupled to an external energy source.[22]

MAKING LIFE IN A TEST TUBE

Many scientists believe we will soon be able to make life ourselves, in the laboratory. In a limited sense, it has already been done. In 2002 a team at the State University of New York, Stony Brook, was able to assemble a polio virus from scratch, using commercially available molecular building blocks. But a virus is not a fully autonomous organism (it cannot reproduce on its own). Bacteria are, and Hamilton Smith and his colleagues at the J. Craig Venter Institute in California have assembled an entire synthetic bacterial genome of 582,970 base-pairs. They were able to insert it into a host bacterium, but at the time of writing they had yet to coax their customized genome to 'boot up' and do anything. Craig Venter himself has been re-engineering the genetic material of small bacteria to create the simplest autonomous cell. Significant though these advances are, a word of caution is necessary. The latter two experiments do not really count as 'making life'. Rather, they adapt existing organisms, in all their fantastic complexity, to make new types of organisms.

Even if an entire autonomous microbe is eventually built *ab initio* without any use of pre-existing life forms at all, it would still not settle the issue of the cosmic imperative. Life began in nature without the benefit of high-tech laboratories and delicate step-by-step procedures implemented under carefully controlled conditions. Above all, it got going without the use of an intelligent designer such as Craig Venter, setting out with a specific goal in mind. Mother Nature created life in the grubby conditions of a newly formed planet (or somewhere else, we don't know), exploiting natural, random chemical reactions, and with no pre-conceived 'destination life' to guide and shape the reactions. What happened just happened. Quite obviously it is *possible* to make life in the lab – all you have to do is to string together the right molecules in the right way. There is nothing miraculous about it; any difficulty is entirely technical and a matter of garnering sufficient resources; with enough time, money and effort, it could clearly be done. But it won't cast much light on how widespread life is in the universe. If it turned out that there were very many ways to make life in the lab, and not too many carefully controlled steps needed to 'boot it up', it would shorten the odds in favour of the cosmic imperative. But creating a totally synthetic organism wouldn't on its own prove that life is ubiquitous.

Summing up then, the probability of life emerging from non-life can be placed on a spectrum ranging from infinitesimal (Monod's position) to almost inevitable (de Duve's position), or anywhere in between. It is frustrating that so basic and crucial an issue remains imponderable. Can we make any progress at all? Indeed we can. In fact, there is an obvious and direct way to confirm if a cosmic imperative is at work, and that is to find a second sample of life.

SEEKING A SECOND GENESIS ON MARS

Everybody agrees that Mars offers the best current hope for finding life beyond Earth.[23] In 1977, NASA sent to Mars two spacecraft called Viking, with the express purpose of seeking microbial life in the surface dirt. Few people appreciate that Viking remains the only successful mission by any space agency to look for extraterrestrial life. *The only one.* The media tend to present all Mars exploration as part of the search for life, but this is a sly piece of disinformation. It is true that some Mars exploration – looking for water, for example – bears indirectly on the question of life, but explicitly biological experiments have for thirty years been systematically eliminated from NASA missions. The European Space Agency is equally lukewarm about the search for Martian biology. Their Mars Express mission, launched in 2003, included only as a belated afterthought Britain's tiny Beagle 2 module. Built on a shoestring budget and not tested properly because of the rush, Beagle 2 was designed to sniff out life on the Martian surface. Sadly, it disappeared without trace. All we currently have to go on are the results of Viking.

Both Viking spacecraft were equipped with a robot arm and shovel to dig up the fine Martian dust and deliver it to little on-board laboratories where four life detection experiments were performed (see Plate 4). The experiments were designed to be as general as possible within the framework of carbon-based life, as there was no reason to suppose that Mars life and Earth life would be the same. One instrument, with the cumbersome name of gas chromatograph mass spectrometer, was built to detect organic molecules, such as the decomposed detritus of once-living cells. Another looked for several specific gases given off or absorbed by any organisms when in the presence of a nutrient medium. A third sought evidence of photosynthesis. The final experiment was designed to detect carbon uptake by adding a nutrient broth to the dirt and seeing whether anything metabolized it. A positive sign that the broth was being consumed by microbes would be the emission of a carbonaceous gas, such as carbon dioxide or methane. To monitor the gas production, the carbon atoms used in the broth included a radioactive isotope, C^{14}, as a label. For this reason the procedure was called the labelled release, or LR, experiment.

The Viking mission was a huge success, and stands as an immense tribute to NASA. Both spacecraft landed safely in widely separated locations. The robot arms deployed properly, the cameras worked and the on-board experiments went off almost without a hitch, and all using 1960s technology. The results were eagerly awaited by scientists and public alike. I recall being on vacation in the former Yugoslavia when the spacecraft landed, and seeing the banner headlines in English on newsstands in Dubrovnic. After centuries of speculation about life on Mars, the time had come to put the idea to a proper scientific test.

The data sent back by the spacecraft painted a confused picture, unfortunately. The mass spectrometer found no trace of organic material, which was odd, because even if there is no life in the Martian soil, small amounts of organic gunk are delivered from space by comets, and should have shown up. Two more experiments were ambiguous. By contrast, the LR experiment gave a strongly positive result. The broth was hungrily devoured and radioactive carbon dioxide came off as hoped – on both spacecraft. When the mixture was heated to 160°C, the strong reaction ceased, as it would if it had been caused by microbes subsequently killed by the high temperatures. On the face of it, the LR experiment had found life. But that was not NASA's spin. Given the indecisive results of the other three experiments, the overall conclusion was 'no life detected on Mars'. It remains the official position today, and is clearly stated as such on the placard in front of a Viking replica at the Air and Space Museum in Washington, DC. The positive results of LR are attributed by most scientists to highly reactive soils created by the harsh Martian surface environment, and especially the effect of ultraviolet radiation.

The designer of the LR experiment, Gilbert Levin, contests NASA's conclusion. He still maintains he found life on Mars. Today, Gil is a colleague of mine in the Beyond Center at Arizona State University, where he holds the position of Adjunct Professor. Back in the 1970s he anticipated the possibility of an ambiguous result from LR, and had a plan to circumvent it. Nearly all organic molecules possess a definite handedness. For example, DNA is a right-handed spiral; seen in a mirror, the handedness is reversed. The technical term for handedness is 'chirality', and it is believed by most scientists to be a universal feature of life. Known life almost always uses right-handed sugars and left-handed amino acids. The laws of chemistry, though, are mirror-symmetric – they do not favour one chirality over the other. So a great way to tell the difference between biological activity and simple chemistry is to look for chiral discrimination – a reaction favouring one chiral form over the other. Gil wanted to run the LR experiment with two broths, one having left-handed amino acids and right-handed sugars, the other using their mirror forms. Thus, had the Mars soil fizzed equally for both, a simple chemical reaction would be the most likely explanation – the one most scientists now back. But if biology had been responsible, then there would have been a marked difference in response between the two forms of broth.[24] Sadly, this refinement was eliminated for reasons of cost. As a result, the Viking experiments remain an exasperating mystery.

In spite of the definitive 'no life detected' conclusion from Viking, many scientists have in recent years warmed to the idea that there might be life on Mars after all. Or at least, that there might have been life there billions of years ago. This shift in attitude is largely due to the accumulating evidence that Mars once had liquid water in reasonable abundance. Photographs show ancient river valleys and lake beds, and on-the-ground experiments confirm that water has flowed over rocks. Today the water is locked up as polar ice and permafrost, but episodic local or global heating may still occur, e.g. as a result of climatic shifts or comet impacts, enabling liquid water to exist briefly on the surface. Water should also be present deep underground, where the internal heat of the planet maintains temperatures above freezing. Mars also has volcanoes which can cause local heating, and there is even evidence for hydrothermal systems, where geothermal hot spots bring about sustained cycling of water over extended periods. On Earth, ancient hydrothermal systems are associated with the oldest traces of life (in the Pilbara hills for example). Indeed, many astrobiologists think terrestrial life actually began in such a setting. As I mentioned earlier, all the evidence suggests that, three or four billion years ago, Mars was markedly warmer and wetter, presumably as a result of a much thicker atmosphere leading to massive greenhouse warming. The environment at that time would have been suitable for microbes; indeed, some hardy terrestrial bacteria could probably survive under current Martian conditions.

If Mars was, or in a limited sense still is, 'Earth-like', we should be able to find evidence of life there, if it exists (or once existed). It might come from a more refined Viking-type probe, from a mission designed to bring rock samples back to Earth, or from a manned expedition. While life on the harsh surface of Mars remains a long shot, subsurface microbes dwelling in aquifers hundreds of metres underground are distinctly possible. They might betray their presence through waste gases such as methane being exuded to the surface. In the next thirty years, scientists may well find clear evidence that microbes existed on Mars at some stage in the planet's history.

Most people mistakenly leap to the conclusion that the discovery of life on Mars would imply that the universe is seething with it. But things are not that simple. As I explained at the beginning of this chapter, Mars and Earth are not quarantined. They regularly exchange material in the form of ejected rocks, and while the traffic from Mars to Earth greatly exceeds that going the other way, over astronomical history huge quantities of terrestrial material will have landed on Mars, much of it infested with microbes. Most of the passengers will have perished on the journey, but not all. If Mars was long ago more Earth-like than today, at least some of these terrestrial stowaways will have flourished in their new home. Conversely, it is entirely possible that terrestrial life did not start on Earth, but came here from Mars. Either way, the mere fact of finding life on Mars will not in itself be enough to establish the cosmic imperative. One would need to demonstrate that life has started *from scratch* on both Mars and Earth, i.e. in both places independently. The ongoing intermingling of Earth and Mars life by exchanged rocks would at the very least severely complicate the story, making it hard to untangle how and where life began, and whether there was one genesis or two.

What about life beyond the solar system? There is only an infinitesimal chance that a rock blasted off Earth would ever hit another Earth-like planet in another star system, and even if it did, there is little prospect that any microbes would survive for the vast lengths of time needed to get there. So the contamination problem is irrelevant. Detecting signs of life on an extra-solar planet would thus be clear evidence for a second, independent, genesis. Astronomers have ambitious plans for large space-based optical systems that could detect the presence of oxygen and perhaps even photosynthesis on extra-solar planets, but the technical challenges are formidable and unlikely to be solved in the near future.

If we have to rely on satellites and space probes to decide whether or not life is a fluke, we could be in for a very long wait. Fortunately, there is another way to test the cosmic imperative, a way that avoids expensive space missions altogether – a way that until recently has been overlooked. We might just be able to settle the matter without ever leaving Earth. No planet is more Earth-like than Earth itself, so if life really does form readily in Earth-like conditions – as the cosmic imperative demands – then it should have started many times over right here on our home planet.

Perhaps it did.

3

A Shadow Biosphere

A box without hinges, key, or lid, yet golden treasure inside is hid.

J. R. R. Tolkien

SEEKING A SECOND GENESIS ON EARTH

If life started more than once on Earth, we could be virtually certain that the universe is teeming with it. Unless there is something *very* peculiar about our planet, it is inconceivable that life would have begun twice on one Earth-like planet but hardly ever on all the rest. Until recently, biologists generally assumed, without giving it too much thought, that all life on Earth is the *same life*, with every organism that ever lived having descended from a common genesis. But how do we know that is so? Could there be two or more different sorts of life on this planet? Has anybody actually looked?

Here's one plausible scenario for how life might have begun repeatedly. As I mentioned in Chapter 2, for about 700 million years after its formation, Earth was subjected to a remorseless barrage of asteroids and comets, the biggest of which could have sterilized the whole planet. Between big impacts, however, conditions would have been less hostile. These quiescent episodes may have lasted many millions of years. According to the 'cosmic imperative' account of life's origin, which we are seeking to test, the lulls may have lasted long enough for life to get under way. For a while, primitive microbes would thrive and spread, only to be obliterated by the next big impact. Then there would be another lull, and life would start again – and get annihilated once more. The early history of life on Earth may thus have been a long series of stop-go biological 'experiments', with many genesis events in sequence producing many varieties of life, an idea first suggested by two Caltech geologists, Kevin Maher and David Stevenson.[1] Their theory was plausible enough, but at the time they overlooked an important corollary. Each sterilizing impact would have ejected a massive quantity of material into orbit round the sun, conveying with it any micro-organisms that may have been in residence. Some of the ejected rocks would eventually find their way back to Earth after the effects of the impact had faded. Dormant microbes could withstand a space environment for millions of years when cocooned in a rock, so some at least would have returned alive and well and ready to resume normal life. However, in the meantime, while Life I was hanging out in space, Life II had formed during the next lull, and become ensconced. There would now be two forms of life on Earth at the same time. This sequence of events may have happened again and again, so that by the time the heavy bombardment faded, there could have been many different sorts of terrestrial life descended from many different geneses.[2]

The foregoing scenario for multiple origins is by no means the only one. Life may have begun independently at many different geographical locations, perhaps remaining trapped in isolated pockets for eons. Some deep-living microbes, cowering in their subterranean refuge, might have been spared the heat of the bombardment, and surfaced only after another form of life had emerged up above. Or life may have started on Mars many times and come in its various manifestations to Earth sporadically over millions of years. It may even have begun on both Mars and Earth, and been transferred between these planets in impact ejecta, to mingle with the indigenous life on arrival. For the purpose of this chapter, the specifics don't matter. All that concerns us for testing the cosmic imperative is whether life started more than once. If it did, what evidence might there be?

Direct confirmation could come from the discovery of living descendants of other genesis events, sharing our planet with us, and constituting a shadow biosphere.[3] A good way to describe this situation is in terms of the tree of life, which illustrates how life developed more and more branches over time, diversifying through successive speciation

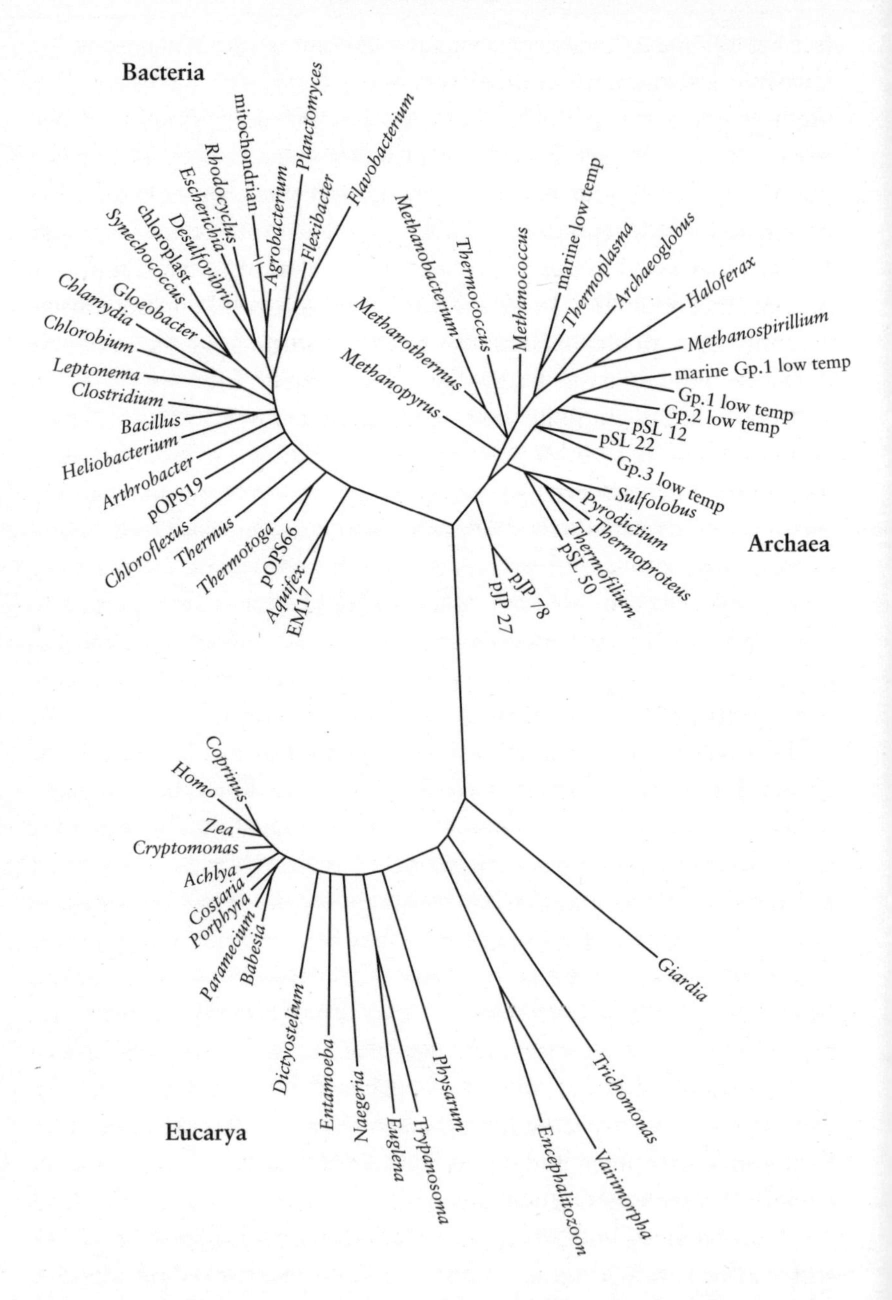

Fig. 2. The tree of life, showing the genetic relatedness of different species. Most species (including all the bacteria and archaea) are microbes. Our species (*Homo*) is shown near the tail of the domain of eucarya.

(see Fig. 2). Life today is represented by millions of different species, but if we trace evolution backwards over billions of years, then they converge on the 'trunk of the tree'. Thus humans and chimpanzees can trace their descent from a common ancestor living in Africa between 7 and 5 million years ago. Go back further, and all mammals converge, then all vertebrates, and so on, to primordial microbes three or four billion years ago. Richard Dawkins has described this biological journey back in time in his engaging book *The Ancestor's Tale*.[4] The question I am then raising is simply, does all life on Earth belong to this *single* tree, or might there in fact be more than one tree? Might there even be a forest?

When I began mulling these ideas over a few years ago,[5] I was amazed to find that nobody had really thought much about evidence for multiple genesis events. Astrobiologists have been busy figuring out how to detect a different form of life on Mars, but it hadn't occurred to many people to hunt for alternative forms of life on our own doorstep. I did, however, find enough open-minded scientists to attend a workshop at Arizona State University in December 2006 and brainstorm a few ideas. The result was a groundbreaking research paper[6] setting out a strategy to 'seek out new forms of life', as the mission statement of *Star Trek* proclaims, not light years out in the galaxy, but on Earth itself.

Before getting into the details, let me summarize why biologists think all *known* life shares a common origin. The main evidence comes from biochemistry and molecular biology. Oak trees, whales, mushrooms and bacteria may look very different, but their internal workings are all organized around the same system. They all use DNA and RNA to store information, and proteins to serve as enzymes and as structural building blocks. Energy is stored and released using molecules known as ATP. Many identical, or at least very similar, genes are found in distinctly different species; for example, humans share 63 per cent of their genes with mice and 38 per cent with yeast. The real clincher comes from the genetic code, the mathematical scheme that translates the data contained in DNA into instructions for making proteins. DNA stores information as sequences of molecular units called nucleotides. There are four different nucleotides, normally labelled by the letters G, C, A and T. What makes you *you* and your dog a dog hinges entirely on the sequence of those letters. (It takes millions of letters to specify you or your dog.) The letters spell out, among other things, the instructions for molecular contraptions called ribosomes to assemble proteins by stringing together amino acids in the correct order. To achieve this specification, known life clusters the nucleotides in DNA into groups of three (for example, AGT). There are sixty-four different possible triplet combinations available to specify the requisite twenty-one different types of amino acids, so choices need to be made about what codes for which. The number of such choices is enormous, because of the huge range of possible permutations, but all known species use the same code.

The fact that such complicated and specific features as ribosomes, ATP and the triplet code are found to be universal would be very hard to explain unless all the species had descended from a universal ancestor – ancient cells that already incorporated those distinctive features. By sequencing genes, it is possible to actually construct a common genetic tree and display the shared descent. Over time, species tend to drift apart genetically, so the number of common genes declines. The slow and cumulative divergence provides a measure for how long ago two given species differentiated. The genetic tree is mirrored in the fossil record, which also charts the steady accumulation of changes and speciation.

Nobody doubts that familiar multicelled organisms lie on the same tree. The animals in the zoo, the plants in your garden, the birds in the sky and the fish in the sea all represent a single type of life. But this is only part of the story: the vast majority of species are microbes. As Stephen Jay Gould so graphically expressed it, 'Our planet has always been in the "Age of Bacteria," ever since the first fossils – bacteria, of course – were entombed in rocks more than 3 billion years ago. On any possible, reasonable or fair criterion, bacteria are—and always have been—the dominant forms of life on Earth.'[7] Under a microscope, many microbes look almost the same – little blobs and rods, sometimes with bits sticking out. You can't tell by looking what goes on inside. If you examine the innards of a microbe, chances are you will find the same stuff – DNA, proteins, ribosomes – as is found in you or me. At least, that has been the experience so far. But microbiologists have only just scratched the surface of the microbial realm. Our world is literally seething with these tiny organisms. Just one cubic centimetre of soil might contain millions of different species adding up to billions of microbes in all, and the vast majority haven't even been classified, let alone analysed. Nobody knows for sure what they are; for all we know, some of them could be life as we do not know it.

To investigate a species of microbe fully, you first need to culture it in the laboratory and then study its biochemistry, e.g. by sequencing its genome to position it on the tree. This technique, whilst undoubtedly important, has its problems. Many microbes don't like being plucked out of their natural habitat and cannot be cultured easily. Some resist gene sequencing. Because the chemical techniques used to analyse microbes are customized and targeted to life as we know it, they wouldn't work on an alternative form of biology. Should there be a different type of microbial life out there, it is very likely to be overlooked, simply because it would be unresponsive to the biochemists' probes used so far. In a laboratory sample it might well get thrown out with the garbage. If you set out to study life as we know it, then what you find will inevitably be life as we know it. It's therefore an open question whether some microbes might actually be the descendants of a different genesis.

WEIRD EXTREMOPHILES

How might we go about identifying life as we *don't* know it? Given the large measure of chance in evolution, it's highly unlikely that organisms from separate origins would have the same biochemistry. Astrobiologists refer to known organisms as 'standard life' and to the hypothetical alternative forms as 'weird life'. (Weird life could be alien life in the sense of 'not one of us', but also in the sense of having an extraterrestrial, e.g. Martian, origin. As I mentioned above, the distinction isn't important for present purposes.)

Part of the problem in searching for weird life is that we don't know exactly what to look for. One strategy is to look in weird places, keeping an eye open for anything that is living. But how weird is weird? Over the past three decades, biologists have been repeatedly amazed to find life surviving or even thriving in environments previously thought to be utterly lethal. In the 1970s, microbes were discovered inhabiting hot springs such as in the Yellowstone National Park. Some of these hardy organisms can withstand temperatures of 90°C, and for obvious reasons they are called thermophiles. That was amazing enough, but more surprises lay in store. Exploration of volcanic vents on the ocean floor by the submarine *Alvin* revealed entire ecosystems in total darkness, close to 'black smokers' – mineral chimneys in the seabed spewing forth dusky fluid at temperatures up to 350°C (see Plate 6). The primary producers at the base of the food chain are microbes that cluster around the stream of scalding effluent, tolerating temperatures up to, and in some cases exceeding, 120°C. This is well above the normal boiling point of water (the water doesn't actually boil because of the high pressure). These extreme heat-loving microbes are called hyperthermophiles. They survive in the dark because they don't require light for energy. Rather, they metabolize and make biomass directly from gases dissolved in the fluid emanating from the Earth's crust.[8]

Many other species of microbes have been discovered living in different extreme conditions. For example, some organisms, which rejoice in the name of psychrophiles, can tolerate extreme cold – maybe as low as –20°C – before they stop growing. Others can withstand acid strong enough to burn human flesh, while yet others endure equally corrosive alkaline conditions. The Dead Sea turns out to be a misnomer, because it is host to several species of halophiles – organisms that live happily in very high salt concentrations. Perhaps most remarkable of all are radiation-resilient microbes like *Deinococcus radiodurans* (see Plate 5), which can survive such high doses of radiation that they have been found living in the waste pools of nuclear reactors.

Collectively these microbial oddballs are known as 'extremophiles'. Notwithstanding their exotic nature, to date all extremophiles that have been analysed have turned out to be standard life – they belong to the same tree of life as you and me. Their existence proves that the range of conditions under which standard life can survive is much broader than previously suspected. Nevertheless there are limits. All standard life requires liquid water, for example. That alone brackets the temperature and pressure range.

If there is a shadow biosphere, it might be occupied by weird 'hyper-extremophiles' inhabiting environments that lie beyond the reach of even the hardiest form of standard life, and have so far escaped detection because nobody thought to look for any form of life under such extreme conditions. A good example is temperature. Standard hyperthermophiles seem to have an upper limit of about 130°C – and for good reason. The intense heat disrupts vital molecules, and even with a host of repair and protection mechanisms, DNA and proteins start to unravel and disintegrate if they are subjected to temperatures much in excess of 120°C. Suppose we find nothing living between 130°C and 170°C in a deep-ocean volcanic-vent system, but then discover microbes thriving there between 170°C and 200°C? The discontinuity in temperature range would be a strong indicator that we were dealing with weird life as opposed to standard life that had simply pushed the temperature envelope higher.

Another limit is depth. In the 1980s the maverick astrophysicist Thomas Gold of Cornell University supervised an experimental oil-drilling project in Sweden, and created a stir when he claimed to have discovered life at the bottom of a borehole several kilometres deep.[9] Not many people believed him. Within a few years, though, other researchers began finding micro-organisms living in the pores of rocks deep underground. But that was just the start. Rock cores from bore-holes drilled into the seabed were found to contain millions of microbes per cubic centimetre, down as deep as the drills could go (about a kilometre). It soon became clear that there is ample room *inside* our planet for microbial habitation.[10] Nobody knows how extensive this deep, hot biosphere might be, or just how far down it stretches; Gold conjectured that there is as much biomass under the surface as on it. Be that as it may, we can easily imagine many isolated, or nearly isolated, subterranean ecosystems, each self-sustaining, and by and large separated from the regular biosphere.

In fact, three ecosystems have been discovered that are almost completely isolated from the rest of the biosphere.[11] Buried deep underground, these extraordinary microbial communities are examples of hydrogen-powered life. The hydrogen is produced by the dissociation of water coming into contact with hot rocks or, in one case, by radio-activity. The organisms get energy and make biomass by combining the hydrogen with dissolved carbon dioxide, and giving off methane as a waste product.[12] Many of them are thermophiles or hyperthermophiles, because the Earth's crust gets progressively hotter with depth. In spite of their splendid isolation, however, all the occupants of these three subsurface ecosystems turn out to be standard life. But it is clear that scientists have so far glimpsed only the tip of the iceberg. An intriguing question is whether some of these pockets might be inhabited by weird rather than standard life forms. It is entirely possible that a future drilling project, on land or at sea, will hit a pocket of weird life. Even if we don't get lucky and actually penetrate such a pocket, we might still obtain indirect evidence for concealed weird life. For ex-ample, standard life is preyed upon by viruses, mostly without any ill effect.[13] They invade plants, animals and microbes. Because they are so tiny, viruses get conveyed to a much wider range of environments than microbial cells. They are everywhere – in soil, air and water. The ocean is pretty much a case of 'virus soup', with up to 10 billion viral particles per litre of seawater. If weird microorganisms are confined to Earth's subsurface (or anywhere else on Earth for that matter), it is likely that 'weird viruses', adapted to interact with them, will spread themselves around the biosphere. They could be present, maybe only at very low levels, amid regular viruses in seawater or air. As far as I know, nobody has thought to look for them.

There are plenty of other places that could be home for isolated weird extremophiles, places so harsh they lie beyond the comfort zone for standard life. The inner core of the Atacama Desert is one place (see Plate 7). It is so dry and oxidizing, bacteria can't metabolize. NASA has a field station there, but so far there is no evidence for any carbon chemistry that could be attributed to weird life. Other possible locations include the upper atmosphere, cold dry plateaux and mountain tops (where high-UV flux is a problem for standard life), ice deposits at temperatures below –40°C, and lakes heavily contaminated with metals toxic to known life. The technical way of summarizing all this is to envisage a multidimensional 'parameter space' of variables such as temperature, pressure, acidity (pH), salinity, radiation, etc. Life as we know it is confined to a finite region of this parameter space, although discoveries in recent years have pushed the boundaries of the 'habitability region' surprisingly far. Still, there will always be an outer limit. A shadow biosphere that is ecologically separate from the regular biosphere would exist in a disconnected region of parameter space. We don't need to confine our search for weird microbes to a single parameter like temperature; it's possible that some combination such as temperature and acidity together is more relevant.

The challenge is to spot the weird microbes if they are present at very low relative abundance. One idea we are working on at the Beyond Center is to make a variant of Gil Levin's Labelled Release (LR) experiment that went to Mars on Viking. After all, this experiment was designed precisely to find organisms of an unspecified variety, using a very general definition of life that relied only on the ability to cycle carbon through its system, something that we expect shadow life to do. The secret of the LR experiment lies with its astonishing sensitivity. As I explained earlier, it works by providing a nutrient broth tagged with radioactive carbon (C^{14}). Any carbon cycling due to metabolism is detected by looking for C^{14} in emitted carbon dioxide. Because even the tiniest levels of radiation are easy to measure, the LR experiment can register trace amounts of activity. If there are weird bugs out there on high mountaintops, in the core of the Atacama Desert or wherever (and assuming they don't choke on the broth so carefully provided), Gil's experiment could find them. The first step will be to determine whether or not they are just an even more extreme extremophile belonging to the standard tree of life, or descendants of another genesis.[14]

ALIENS AMONG US

In the previous section, I discussed the idea that weird life might be restricted to isolated pockets beyond the reach of standard life, making it easy to spot. Much harder would be if weird life and regular life are intermingled. A persistent science fiction theme is that alien beings are living clandestinely among us, indistinguishable from humans. A classic of its kind was *Quatermass 2*, a BBC television horror series of the 1950s, in which unlucky individuals get 'taken over' by aliens. In others, like the long-running 1960s American television series *The Invaders*, aliens disguised as humans infiltrate our society. The popularity of this genre is in part financial: it's cheaper to use human actors with little or no make-up to play the part of the aliens. For decades it also fed off fears of the Cold War, and the 'reds-under-the-bed' neuroses of many Westerners. Improvements in special effects, costume design and computer-generated imagery finally brought about a shift in the way that aliens were portrayed, so that by the time the movies *Star Wars* and *Alien* were released, alien anatomy had become much more varied and less humanoid.

So much for science fiction. Now it seems that a Lilliputian variant of the alien infiltration theme could actually be true. If weird microbes look like standard bacteria and inhabit the same environment as us, they may have already been spotted, but lacking a visible uniform that proclaims membership of an alternative club they wouldn't have excited comment – they would remain hidden in the microbial crowd.[15] There could literally be alien organisms right under our noses (or even in our noses!), as yet unrecognized for what they are. The thorny problem is how to identify them.

One way is biochemically. Two microbes may look similar yet have very different chemistry going on inside. If we could know in advance what an alternative biochemistry might be, we could then test microbial samples for signs of it. The trick is to guess right. As we don't know precisely what we are looking for, this is quite a challenge. But we can make some educated guesses. An obvious example is chirality – the selection of right-handed sugars and left-handed amino acids rather than their mirror images (see p. 39). If life were to start over again, there is a chance it would choose the opposite handedness next time (see Fig. 3). Even if this 'mirror' life resembled standard life in all other respects (for example, by using the same nucleic acids and proteins), it would stand out – not visually, but biochemically. What is needed is a chemical filter to target standard life, but not mirror life. I was discussing this problem with my wife Pauline a few years ago, when she came up with a bright idea of what to do. Surely, she suggested, mirror life would turn up its proverbial nose at a culture medium that is tasty to standard life, but would gobble up 'mirror soup' – a medium in which standard sugars and amino acids are replaced by their mirror images. For standard life, it would be vice versa. By this means one might sort out the sheep from the goats. We persuaded Richard Hoover and Elena Pikuta to perform a pilot mirror

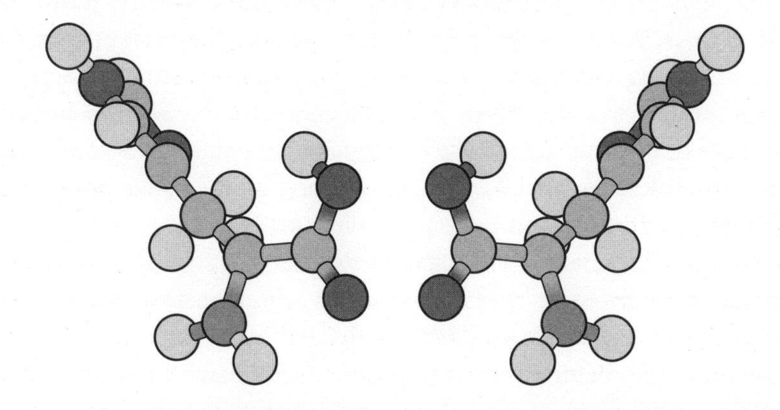

Fig. 3. Life and mirror life. If all the molecules standard life uses (like this amino acid) were replaced by their mirror images, the result would be an organism that would require 'mirror' food.

soup experiment at NASA's Marshall Spaceflight Center in Huntsville, Alabama. The results were very curious. Hoover and Pikuta discovered a novel extremophile from a highly alkaline lake in California that ate the mirror soup with gusto. They named it *aerovirgula multivorans* (meaning, roughly, unfussy little goat).[16] Sadly, this was not the mirror microbe we had hoped for, but a standard microbe cleverly adapted to cope with mirror food. It turns out that standard life sometimes makes use of mirror molecules (for example in cell membranes), and some standard microbes are loaded with enzymes that can chop up molecules of the 'wrong' handedness and turn them into useful products. According to Hoover, *aerovirgula multivorans* was able to grow by digesting a mirror version of the sugar arabinose, but *couldn't* grow using standard arabinose, which is surprising. So the chirality story is a bit perplexing and clearly more complicated than we originally envisaged. Nevertheless, using chirality as a signature for weird life remains an obvious and easy technique.

Another clue could come from the building blocks that weird life might use. As I mentioned, standard life uses twenty-one types of amino acids to make proteins, but many other varieties exist. In 1969 an unusual meteorite fell near the town of Murchison in Australia, belonging to a rare class known as carbonaceous chondrites (see Plate 8). The Murchison meteorite contains abundant organic material – so abundant it smells of petrol – including many amino acids that standard life doesn't use. A few people have jumped to the conclusion that the meteorite was once inhabited by alien microbes that decomposed, leaving their exotic amino acid contents for us to find among the corpses. But this conclusion is a stretch; it's more likely that these organic molecules formed somewhere in space. As I mentioned in Chapter 2, it's not hard to make amino acids in the laboratory, so presumably there are many natural ways for them to form too. The early Earth may have been coated with carbonaceous material from meteorites and interplanetary grains that fell like manna from heaven, providing raw materials from which the first life may have emerged. If this is correct, the original cells would have been able to pick and choose from the organic cocktail. To the best of our knowledge, the twenty-one chosen by known life do not constitute a unique set; other choices could have been made, and maybe *were* made if life started many times.

Steve Benner is a biochemist and a world expert on synthetic biology. He knows a lot about how to make cells that incorporate 'unnatural' components that he himself inserts.[17] One component shunned by regular life, but which Benner thinks is good for synthetic life, is a class of molecules known as 2-methylamino acids. If we found organisms employing these amino acids, it would be a strong indicator of something new and weird. In fact, we wouldn't need to spot the microbes themselves: organic detritus containing 2-methylamino acids, especially if it displayed a preferred chirality, would be a tell-tale sign. Benner's suggestion for amino acids is part of a general strategy: make a list of organic molecules that known life *doesn't* make, which are not breakdown products of known life, and preferably don't form naturally by non-biological processes. Then just go out and look for them. Nobody has yet tried this: there has been no systematic survey for weird organics in the environment.

Related to the issue of amino acids is the genetic code, which, as I explained earlier, is universal for standard life. We can imagine an alternative type of life made up of DNA and the *same* suite of twenty-one amino acids, but employing a different genetic code. It would be easy to overlook organisms with this 'near miss' biochemistry, yet they would betray themselves readily if studied in detail by molecular biologists. More likely, if weird life started from scratch independently of standard life, it would use a *different* set of amino acids, so it would also have to employ a different genetic code. We can even imagine life in which two of the four nucleotides G, C, A and T are absent, or replaced by a different nucleotide, or in which there are more nucleotides (six instead of four, say). These are all candidates for synthetic life, and therefore are also possibilities for alternative forms of natural life. Because there is little chance that micro-organisms using fundamentally different biochemistry would respond meaningfully to standard biochemical techniques, weird microbes of this sort might be all around us, so far unidentified.

A more radical form of weird life would be organisms that use different chemical elements. Life as we know it is based on the unique properties of carbon chemistry, but it also uses several other key elements, specifically, hydrogen (H), nitrogen (N), oxygen (O), phosphorus (P) and sulphur (S). There has been some speculation that silicon could substitute for carbon, a conjecture that got as far as an episode of *Star Trek*, but hasn't been pursued very seriously by biochemists because silicon can't form the extraordinary range of complex molecules that carbon can. A more plausible candidate came from my collaborator Felisa Wolfe-Simon, who suggested that phosphorus could be replaced by arsenic.[18] Arsenic can do the same structural and energy-storage jobs as phosphorus, but it can go one better, by providing an energy (i.e. food) source too.[19] In fact, there are microbes that exploit arsenic, but they don't inhale it, so to speak: the arsenic compound gets stripped of its energy and the arsenic is then summarily expelled. Arsenic is a poison precisely because our bodies have a hard time telling it apart from phosphorus. Felisa hopes to find weird microbes with arsenic incorporated in their vitals, and for which phosphorus would be the poison.

HOW TO TELL A ROOT FROM A BRANCH

If weird life is discovered, the first priority will be to determine whether it belongs to a genuinely separate tree of life, or is merely a hitherto undiscovered branch on the known tree of life. The distinction is depicted in Fig. 4. Suppose we are presented with two radically different forms of life, which we are tempted to attribute to separate trees, each with an independent origin (by which I mean independent transitions from non-life to life), as shown in Fig. 4a. On further investigation, however, we may find that 'below ground' the two trunks join in a common root system (Fig. 4b): that is, the different forms of life belong on a single tree after all, but they branched apart before the last common ancestor of all standard life.

The known tree of life consists of three distinct 'bushes' that branched apart billions of years ago (see Fig. 6). One bush contains the bacteria. Another has all multicellular life, from humans to hedgehogs. It also has complex single-celled organisms like the amoeba. This is the domain of 'eucarya.' The third bush consists solely of microbes, but they are as different from bacteria as they are from us, and have the collective name 'archaea'. The question I am raising is how do we know that there isn't a *fourth* bush, waiting to be discovered, that split away even earlier than the trifurcation into bacteria, eucarya and archaea? If we ever found a new exotic form of life, we would need to eliminate the 'fourth bush' explanation before concluding that it provides evidence for a second tree.

How can a low-lying branch be distinguished from a separate tree? The answer would depend in part on just how weird the weird life is. To use a well-worn phrase, the devil would be in the details. Consider the case of 'mirror life' (organisms with reversed chirality). Is it conceivable that the earliest forms of life were achiral, i.e. built out of mirror-symmetric molecules only, following which the tree split into two domains, one populated by organisms with left-handed sugars and right-handed amino acids, and the other populated by organisms with the mirror forms? This seems extremely unlikely. Small, simple molecules are often mirror symmetric, but molecules of even moderate complexity necessarily have both left- and right-handed versions. It is very doubtful if a system with the complexity of a living organism could arise using only simple achiral molecules. So the discovery of mirror life would be a strong indicator of multiple genesis events.

By contrast, if a form of weird life were discovered that resembled standard life in all but the genetic code it would be easy to argue that

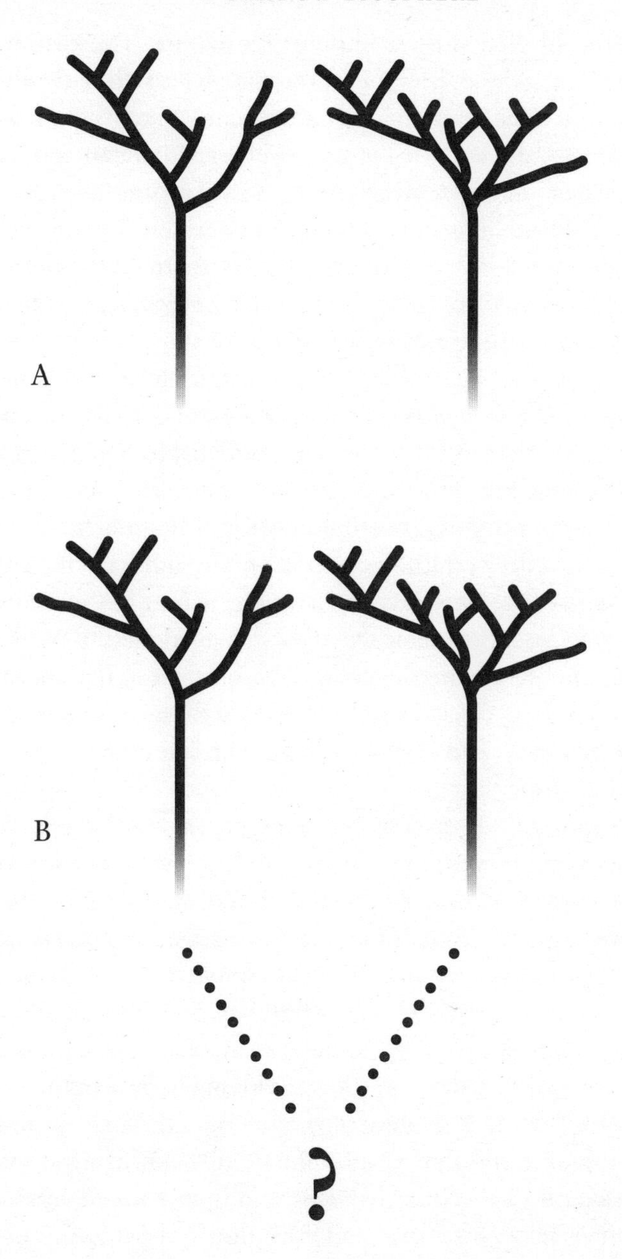

Fig. 4. Tree or forest? If two forms of life exist on Earth, it will be important to determine whether they represent distinct trees of life with independent, separate origins, as in (a), or have merely evolved a long way apart from a common single origin, as in (b).

the two forms of life had a common genesis and a common precursor code, following which life split into two forms that evolved different codes. At least one version of this scenario seems plausible. The triplet code used by known life is complicated, and some biologists have speculated that it evolved from a simpler precursor, perhaps a doublet code based on only two nucleotides (G and C) and ten amino acids. This slimmed-down version of standard life would presumably be less complex, but may have been entirely successful three or four billion years ago. The triplet code might have evolved later, bestowing greater versatility that enabled life to spread to a wider range of environments. The transition from doublet to triplet code may have happened more than once, or the original triplet code could have subsequently split into variants.

An even more intriguing possibility arises. Might some of the 'old-fashioned guys' still be out there, living an ancient lifestyle using only a doublet G-C code? Once again, these 'living fossils' would be overlooked by standard biochemical analysis, but they would be identified clearly enough if researchers chose to look for them.[20] In a similar vein, if arsenic life shows up, we would need to find out whether standard life began that way and then evolved to replace arsenic with phosphorus. Fascinating though the discovery of such precursor organisms may be, it wouldn't get to the real heart of the matter, which is the possibility of multiple origins. To be sure that any weird life really is descended from a second genesis, it would have to be sufficiently different from standard life for no plausible common ancestor to have existed. That criterion would be hard to establish if the two biospheres overlap and use a lot of common chemistry. Still harder would be if the two forms became partially integrated biochemically, e.g. by swapping genes or other structures, thus muddying their separate lineages and confusing the whole evolutionary story. We can't rule out one form of life 'taking over' another, *Quatermass*-like, by infusing key components of itself into a receptive host, especially if two separate forms of life found themselves on convergent evolutionary tracks. All this would be an unwelcome complication. It would be sad and annoying if life started on Earth many times over, but converged and merged, so that we had no hope of untangling its multiple roots.[21] Personally, however, I do not believe evolutionary convergence could ever be that strong. It may throw up similar gross features, but to zero in on a specific biochemical scheme seems very unlikely.

It is often argued that if two different forms of life found themselves side by side, one would eventually gain an advantage and eliminate the other. I have never been convinced that things have to unfold that way. Peaceful coexistence is another possibility, and could arise in two ways. First, if the two forms are sufficiently dissimilar as to be totally indifferent to each other, they wouldn't compete anyway. For example, mirror life would not be in direct competition with known life, because the two forms would mostly use different molecules for food. One form might gain the upper hand in strict numerical terms, but so what? Microbiologists are familiar with the fact that some species are very rare, yet they remain a stable component in the overall microbial population. The second type of peaceful coexistence is where populations of very different sorts of microbes reach an accommodation. The side-by-side cohabitation of bacteria and archaea, two great microbial domains representing millions of species that often share similar niches, provides one example. You might suppose that this tolerance was due to the two domains becoming biochemically integrated – i.e. marriage rather than rivalry. Gene swapping goes on all the time in life, especially among microbes. But in fact archaea and bacteria seem to have jealously guarded certain very basic genes. So far as we know, archaea have never shared with bacteria (or eucarya) their ability to metabolize by making methane, yet methanogenesis is widespread among archaea, occurring in locations as diverse as deep-ocean vents and the human gut. Conversely, photosynthesis has apparently never passed from bacteria (or eucarya) to archaea.[22] So it is clear that very different forms of microbes can compete in the same space for many of the same resources, without one form ever eliminating the other.

Even if the descendants of other origins did go extinct long ago, they could still leave some remnant of their erstwhile existence in the form of ancient fossils and distinctive molecular biomarkers. For example, steranes (molecules with four rings) are produced by complex cells, and are not known to form by any abiotic means. Steranes have been found in trace quantities in microfossils dating back 2.7 billion years. If fossils containing 'mirror' steranes, i.e. of the opposite chirality, were discovered, it could be evidence for ancient mirror life. Many other complex organic molecules from a radically alternative biochemical scheme might survive in rocks for a long time. An indirect way in which extinct weird life might leave a trace is through mineral processing. Many mineral deposits, including iron, copper and gold, are thought to be biogenic – that is, their deposition and concentration have been caused at least in part by the activities of microbes that use these metals for metabolism. A mineral deposit that was impossible for known life to create, yet showed the hallmarks of being biogenic, would be circumstantial evidence for alternative biochemistry at work.

HAS SHADOW LIFE ALREADY BEEN FOUND?

From July to September 2001, the southern part of the Indian state of Kerala was repeatedly soaked by mysterious red-coloured rain. Samples were collected and sent for analysis to Indian and British laboratories. The water was found to contain motile cells resembling bacteria. Before long there were claims that the red rain of Kerala contained extraterrestrial microbes. I was sent some video sequences by Indian researchers that show cells jiggling about, but they are indistinct and could be anything. As so often in these scientific mysteries, the research petered out and the findings remain inconclusive. Several physical mechanisms might explain coloured rain, which turns out to be a persistent feature in Southern India, so the claim that some sort of weird life from space descended on Kerala shouldn't be taken too seriously. On the other hand, if weird UV-tolerant microbes inhabit the very high atmosphere, then we might expect that from time to time meteorological changes would drive them to lower altitudes, whereupon they could nucleate raindrops and ride to the ground. Interestingly, air-dwelling bacteria have been found that nucleate ice crystals by secreting special enzymes, giving them a clever way to reach the ground in snowflakes.[23]

Another intriguing phenomenon is the strange rock coating, found in most of the world's arid zones, known as desert varnish or desert crust. Its origin has been something of a puzzle since Darwin himself remarked on it. The coating certainly contains microbial life, and also unusual combinations of minerals (as a matter of fact, some contain arsenic). The chemical composition of the coating is very different from that of the host rocks. It is not clear whether the varnish is a product of life, or a complex mineral layer that has been invaded by life opportunistically. It does, however, provide a readily accessible source of 'moderately weird' material that merits further study. My colleagues at the Beyond Center carried out a pilot investigation, but so far there has been no follow-up. We are now getting ready to analyse new samples.

Probably the most persistent claim that weird life has already been discovered concerns tiny forms known as nanobacteria. These little blobs measure only a few hundred nanometres across (a nanometre is one billionth of a metre). They resemble bacteria but are too small to contain ribosomes, the protein-making machines that are a key component of all life as we know it. Nanobacteria have been reported in rocks,[24] oil wells[25] and blood.[26] They have been implicated in numerous diseases, ranging from renal disorders to Alzheimer's, and have even attracted the attention of pharmaceutical companies. The claim that these little structures are living organisms, as implied in the use of the term 'bacteria', is highly controversial; if they are, it's hard to see how they could be standard life. They might be a weird form of life that assembles proteins in a novel way, or uses some other type of enzyme. Or they might not be living at all. One theory, suggested by Steve Benner, is that some nanobacteria might be a form of RNA-based life that doesn't need ribosome-made proteins because RNA does the job of both proteins and DNA.[27]

Nanobacteria were propelled to fame by an unlikely figure: President Bill Clinton. In August 1996, Clinton announced that NASA scientists had evidence for life on Mars, in the form of microscopic features inside a meteorite found in Antarctica in 1984, and subsequently shown to have originated on Mars (see Plate 9). The shapes looked for all the world like fossilized bacteria, except they were about ten times smaller than the smallest terrestrial microbes. Some commentators jumped to the conclusion that nanobacteria come from Mars. Many scientists started to believe that living microbes could relocate from Mars to Earth inside meteorites. Everyone was excited. Today the fuss has died down, and extensive analysis of the meteorite has chipped away at the claim that it contains fossilized Martians, to the point where very few scientists continue to believe it.[28]

Whatever the evidence for life in the Mars meteorite, the claim that there are nanobacteria on Earth remains unresolved. Several years ago I visited Philippa Uwins at the University of Queensland in Brisbane, Australia. Philippa had found funny little bacteria-like shapes in samples from an oil-drilling project off the coast of Western Australia, whilst doing a routine analysis for the drilling company. She made her discovery using an electron microscope to study the fine details of the material, and called the shapes by the more neutral name of 'nanobes' (see Plate 11). Like nanobacteria, nanobes are too small to be conventional living cells. Philippa was justifiably thrilled when she detected DNA in her nanobes. She showed me the evidence. Using a type of chemical mixture called a gold colloid, she was able to get the gold to bind to DNA and then, in the microscope images, she could see it was located *inside* the nanobes and not floating loose. That was important, because fragments of DNA from decomposed standard microbes could become stuck to mineral surfaces and preserved. The fact that the nanobes contained DNA suggested to Philippa that they were at least once-living cells, if not still alive, but presumably lacking ribosomes for protein assembly on account of their minuscule dimensions. She was unable to obtain a meaningful DNA sequence, however, which could mean she was dealing with weird DNA-based life that uses a different genetic code. A more prosaic explanation is that nanobes are mineral capsules that have formed around DNA detritus floating in the oily environment.

Research by John Young and his student Jan Martel at the Rockefeller University has led them to conclude that nanobacteria, or nanobes, aren't in fact alive. Young and Martel suggest they are instead chemical complexes made up of organic material combined with common calcium carbonate (limestone), forming amorphous shapes superficially resembling diminutive cells.[29] The researchers are keen to point out that, even so, nanobacteria are not unconnected with the topic of life's origin, because they provide a natural example of chemical self-assembly – a step on the road to life perhaps, even if the nanobacteria are not themselves alive. They draw a comparison with prions – protein-like chemicals that can become malformed in a type of chain reaction, giving rise to illnesses such as kuru and 'mad cow disease'.

The foregoing examples are suggestive, but as yet inconclusive, and obviously require closer investigation. Meanwhile, the hunt for shadow life, or weird life, is picking up pace around the world.

TARGETING THE SHADOW WORLD

As I explained earlier, my colleague Felisa Wolfe-Simon has a hunch there could be weird microbes based on arsenic, and NASA is currently funding a project for her to go look. Where might these arsenophiles lurk? One obvious place is an environment rich in arsenic. Many lakes and springs around the world are arsenic-contaminated and pose a health hazard. Mono Lake in California, an ecological marvel in the eastern Sierra close to the Yosemite National Park, is a picturesque haven for exotic wild life, and none is more exotic than the microbial inhabitants. The lake has exceptionally high arsenic concentration, and is home to many peculiar organisms, some of which seem to use the abundant arsenic to their advantage. The great expert on Mono Lake's arsenophiles is Ron Oremland of the US Geological Survey in Menlo Park, who is hosting the project. To date, none of the microbes he has studied is an authentically weird form of life, with arsenic incorporated in its innards, as Felisa has suggested. Rather, they are all simply unusual adaptations of standard life. But the search for arsenic life has only just begun, and Ron and Felisa have devised a way to speed it up. Samples from the mud at the base of the lake are taken to the laboratory for culturing and experimentation (see Plate 10). There the micro-organisms are subjected to ever-increasing levels of arsenic. In Mono Lake, standard microbes may have adapted to handle arsenic, but their tolerance does have its limits, and at some level of concentration the cells overdose, dying quietly of arsenic poisoning like tiny victims in an Agatha Christie novel. Genuinely arsenic life, by contrast, will lap up the cocktail and thrive. By performing successive culturing operations at higher and higher levels of arsenic concentration, the experimenters expect that any arsenic-based microbes, even if initially present in only trace amounts, will soon out-multiply the standard-life competition, and so come to dominate the microbial population.

A giveaway for arsenic life would be the presence of a structure that is familiar from standard life, but modified by arsenic substituting for phosphorus. One example would be nucleotides – the building blocks of DNA, in which phosphorus plays a central role. Another is in the cell membrane, which is made of a substance called a lipid that contains phosphorus. Both these structures can be explored for signs of arsenic using standard chemical analysis. A third experiment uses radioactive arsenic as a tracer, to see whether it gets incorporated into the biomass.

Another approach we are developing is to sample life as widely as possible from the oceans. In 2004, Craig Venter, having helped sequence the human genome, stunned the scientific world once again when he announced he had isolated a staggering 1.2 million new genes and 1,800 previously unidentified microbes in a sample of water taken from the apparently barren Sargasso Sea. In a telling comment, Venter said, 'We're looking for life on Mars, and we don't even know what's on Earth.'[30] Precisely. Most of what we know about biodiversity in the microbial domain comes from studying the tiny fraction of organisms that can be cultivated in the lab. That is obviously highly unrepresentative. There are certain to be an immense number of rare micro-organisms that have been completely missed by standard molecular methods, perhaps including weird microbes that would in any case fail to respond to standard techniques even at high relative abundance. Venter's so-called shotgun analysis, in which DNA from many cell samples is shattered randomly into bite-sized fragments and then sequenced, enables scientists to measure the genetic diversity within the samples en masse, without the need to separately identify and culture each individually captured species. The challenge is to extend those techniques to pick up any *non*-standard micro-organisms, too, that might constitute part of a shadow biosphere. Ideally this should include weird viruses, or other ultra-small molecular parasites that might be totally novel.

Several ocean sampling projects are now under way, providing a golden opportunity to discover any weird life that may be lurking in the sea. A three-year international project called Tara-Oceans is performing a global sampling exercise, primarily directed at studying the impact of carbon dioxide accumulation on marine biodiversity.

The project will also look at deep-ocean ecosystems and sample microbiology from all the world's oceans. The project's scientists will be on the lookout for a shadow biosphere too, deploying a range of techniques for identifying weird life, and returning selected samples to the Beyond Center for laboratory analysis.

The discovery of a form of life that could have arisen only via a second genesis would be the most sensational event in the history of biology, with sweeping consequences for science and technology. It would also have immediate implications for astrobiology, as we could then be sure that the universe really is teeming with life, as so many commentators glibly assert. However, the goal of SETI is to find not just life, but *intelligent* life beyond Earth. It could be that life is common, but intelligence is rare. What are the chances that, once life gets going on a planet, intelligence will sooner or later evolve?

4

How Much Intelligence is Out There?

Sometimes I think the surest sign that intelligent life exists elsewhere in the universe is that none of it has tried to contact us.

Bill Watterson, cartoonist

PLANET OF THE APES FALLACY

If you could climb aboard a time machine and visit Earth 3.5 billion years ago you would find barren continents and deserted oceans. The only sign of life would be some unexceptional leathery mounds dotted about in tidal shallows. These dome-shaped structures, called stromatolites, vary in size from a few centimetres to a metre. Stromatolites are not themselves living organisms; rather, they comprise mineral layers deposited by microbes inhabiting the structure's surface. As far as we know, there wasn't much else going on 3.5 billion years ago, biologically speaking.

Today, our planet abounds with life. There are millions of species of complex organisms, flying, crawling, burrowing, swimming and photosynthesizing. This rich and elaborate web of life has evolved, sometimes steadily, sometimes in fits and starts, over the billions of years since the age of stromatolites. If a single word is invoked to describe this transformation it is 'progress'. Some people might prefer 'advancement'. The overwhelming impression one gets from studying the evolutionary record is one of biological exuberance, with life spreading almost everywhere, ceaselessly experimenting with new and better adaptations, and exploring ever more complex body plans. In Darwin's eloquent prose, 'Whilst this planet has been cycling on according to the fixed law of gravity, from so simple a beginning, forms most wonderful have been and are being evolved.'[1]

Many biologists (including Darwin himself) loosely endorsed this view of overall evolutionary advancement – a steady onward march from the primitive to the sophisticated, from the simple to the complex. And the pinnacle of that advancement is – you've guessed it – Man. Distinguished by his massive brain and superior intelligence, *Homo sapiens* stands as the archetypal symbol of nature striving towards better, more refined, forms of life. And (so the argument goes) this relentless march of progress is surely not a mere terrestrial aberration, but must be a basic property of the natural order of things, so that we might expect it to be repeated on all planets that support biology. Seed a planet with life, come back a few billion years later, and expect to find culture, language, technology, science and – with luck – radio telescopes. In other words, intelligence, and its manifestation as technological society, is something almost bound to emerge sooner or later, once life gets going, and barring any unfortunate accidents (like the host star blowing up). It is a widespread view, and the one that Carl Sagan, and most other SETI researchers, have taken. But is it right?

The optimistic, or 'progressive', account of intelligence is bolstered by a study of the evolution of brains. Absolute brain size is not itself a good measure of intelligence, because a lot of the brain is used for running the body: big bodies demand big brains. For example, a pussy cat, which has a brain the size of a walnut, is not obviously more stupid than a Bengal tiger. The so-called encephalization quotient (EQ) is an attempt to get around this by comparing the actual brain size with an average or expected brain size for the particular body size of the animal concerned.[2] The reference ratio is taken as 1, so that scores higher than 1 are 'big-brained', lower than 1 'small-brained'. We brainy humans boast an EQ of about 7.5, chimpanzees (our nearest living relatives) 2.5 and dolphins 5.3. (For those who are interested, pussy cats come in at a mediocre 1.[3]) Neanderthals, who were probably not our direct ancestors, but a different branch of the genus *Homo*, had an EQ of about 5.6. If you plot how EQ has evolved in our lineage over time for the past few million years, it seems to show an accelerating trend. Some even claim an exponential rate of growth.[4] It's almost as if intelligence 'took off' as a great evolutionary idea and surged ahead, suggesting that evolution somehow 'favours' it, and will presumably do so on any planet that has organisms with something like a central nervous system.

If only it were that simple. Unfortunately, the popular view of evolution as progress is at best a serious oversimplification, at worst just plain wrong. It is the essence of Darwinism that life cannot 'look ahead' and tailor evolutionary changes to a desirable goal or future opportunity. Mutations occur randomly and will be selected simply on the basis of what works best at the time. Nature cannot foresee the future any more than we can, so the idea that life is actively striving for, or channelled towards, some pre-determined end, is wrong. This point was much stressed by the late Stephen Jay Gould, who used the analogy of a drunk leaning against a wall, who is later found to be lying in the gutter. Did the drunk aim for the gutter? No, he just staggered about at random, but because the wall prevented him moving in the direction away from the gutter, sooner or later he was bound to encounter the kerb, and topple over. The process creates an illusion of directionality due to the asymmetry of the set-up. In the same way, said Gould, life is not *aiming* for complexity or 'advancement'. It starts out simple (of necessity), and there is nowhere to go but up.[5] Life becomes more complex on average over time, not because it is subtly directed towards complexity, but merely because it is randomly exploring the range of possibilities, most of which are more complex than the starting state. Gould believed that the 'progressive' misconception is exacerbated by the metaphor of the tree of life first used by Darwin, which has a clear direction (up), whereas a bush would be a more fitting metaphor. Summarizing this viewpoint, one might say that life simply 'makes it up as it goes along'. And intelligence is just one of those things it made up. What we want to know for SETI, of course, is just how *likely* it is that life will blindly 'blunder into' intelligence (like the drunk), along the evolutionary way. Will it happen a little? A lot? Almost never?

A key factor in addressing these questions is the phenomenon of evolutionary convergence.[6] It occurs when the same biological solution is discovered for a similar problem, but via different routes and from different starting points. Examples abound. Wings have been invented many times – in insects, birds, mammals and even fish. They have arisen independently because flying or gliding has obvious evolutionary advantages in some circumstances, and growing wings by adapting different organs (skin between limbs for flying foxes, fins for fish . . .) is a relatively straightforward step. Eyes have arisen many times too. In fact, there are many different sorts of eyes. Sight also has great advantages, and it is no surprise that evolution has discovered it, independently, again and again.

An interesting debate in biology concerns what general patterns or trends are manifested by evolutionary convergence, and whether it is legitimate to describe some of them in terms of 'available niches'. Let me give an example. With the breakup of the supercontinents Gondwana and Laurasia, animal evolution diverged between the separated continents. What is now Australia finally split away from Gondwana about 50 million years ago and became dominated by marsupials, whereas the other continents became dominated by placental mammals. When the Aborigines reached Australia about 50,000 years ago, they discovered a fierce carnivorous predator, named *Thylacoleo*. Sadly, the *Thylacoleo* is now extinct, possibly from hunting or climate change. This creature evolved from plant-eating marsupials, but ended up looking, eating and behaving very much like the sabre-toothed tiger of North America, which descended from mammalian placental carnivores. Thus, the *Thylacoleo* could be said to have 'occupied the tiger niche' in the Australian ecosystem. This blunt way of putting it implies that there actually is a 'tiger niche' out there, waiting to be filled, just as there is a wing niche and an eye niche.

Because evolutionary convergence is so widespread and powerful, the niche metaphor has some force. But it must be used with great care. What we want to know for SETI is whether there is an 'intelligence niche', which on Earth humans obligingly filled, starting a few million years ago in Africa when our ancestors first walked upright and began using tools – a train of development that led all the way to radio telescopes. And if that reasoning is sound, might we also expect ET to similarly put the 'I' in SETI for us? There is no consensus on the answer. Charley Lineweaver, an astrobiologist at the Australian National University, is highly sceptical of the intelligence niche argument.[7] He likes to compare wings and eyes with trunks. A large African elephant that understood biology might erroneously conclude that 3.5 billion years of evolution was in fact directed towards longer and more versatile trunks, arguing that there is a 'trunk niche' which it, *Loxodonta africana*, has been called upon by Mother Nature to fill. In examining its evolutionary lineage, the elephant might be moved to dwell on a 'nasalization quotient,' (rather than an encephalization quotient). The fossil record would show an evolutionary trail of smaller-trunked predecessors leading (inch by trunk-inch) up to the modern elephant, a trend that might prompt a chauvinistic animal to conclude that because the nasalization quotient had accelerated with time, the magnificently trunked African elephant was truly destined to be.

The ridiculous nature of this line of reasoning is stark when it comes to trunks, but still convinces many people when applied to intelligence. Trunks are, after all, trivial appendages that have had very little impact on the world, whereas human intelligence has reshaped the planet. Is high intelligence not more profound, biologically basic, and generally more significant than long trunks? Well, we would say that, wouldn't we, retorts Lineweaver. We value big brains because that's what we have. Elephants (presumably) value big trunks because that's what they have. There is no objective reason why one is more important, or 'more predestined' than the other. We might just as well expect big-trunked aliens as intelligent aliens, he says. (Amusingly, a 1985 novel by Larry Niven and Jerry Pournelle called *Footfall* does indeed feature elephantesque aliens, who also have the benefit of high intelligence, though not high enough to win a war against us wily humans.) Lineweaver likes to cite the rather dreadful Hollywood movie *Planet of the Apes*, starring Charlton Heston, as a classic example of the purported fallacy. In the movie, humanity is destroyed by nuclear war, but the apes are waiting in the evolutionary wings to occupy the suddenly vacated 'intelligence niche'. Within a few centuries they have 'taken over', and discovered guns, jails and horseback riding, moving a rung up the evolutionary ladder from which *Homo sapiens* has been abruptly displaced.

In the context of SETI, what it boils down to is this: we can make a list of traits, like eyes, wings and perhaps tigerness, for which there do seem to be 'niches waiting', and others like peacock feathers and elephants' trunks that seem to be incidental – even outlandish – accidents of evolution, accidents that are so highly specialized they are unlikely to crop up often. We need to know to which list intelligence belongs. One approach is to ask how long nature took to discover intelligence. The answer is, a very long time compared to eyes and wings. Intelligence could have evolved at any time in the last 300 million years, since the rise of animals, but advanced intelligence (approaching the radio-telescope-building variety) appeared only within the last few hundred thousand years. If there really is 'an intelligence niche' out there it had its chance to be filled with the dinosaurs – otherwise successful creatures who famously 'ruled the Earth' for 200 million years before being wiped out by a comet impact, thus 'paving the way' for mammals. Why didn't dinosaurs evolve big brains, build rockets and fly to the Moon? Chris McKay has addressed this issue: 'It is now considered that the dinosaurs were not the lumbering clods of urban myth but that they were biochemically and behaviorally as sophisticated as present mammals.'[8] If intelligence has such good survival value, why didn't dinosaurs evolve it? They had plenty of time to do so. McKay points out that the small dinosaur *Stenonychosaurus* (now redesignated *Troodon*) had an EQ comparable to that of an octopus (a very smart animal), and was walking the Earth 12 million years before Dinosaur Doomsday. That's longer than the time it has taken human intelligence to evolve from a similar EQ starting point.

Many scientists assert that life on Earth is a single experiment, and one can't conclude much from a solitary evolutionary narrative. But the dinosaur example suggests that evolution has actually had at least *two* chances to do intelligence. In fact, it can be argued that the intelligence experiment has been run *several* times on Earth. Lineweaver has pointed out that no intelligent marsupials evolved in Australia even after 50 million years of physical isolation. Neither did intelligence emerge in South or North America, or in Madagascar, all large and richly populated regions which were separated for much longer than the time it took to produce the human brain. If big brains and intelligence were likely to evolve, surely it would have happened more than once on Earth? Sometimes it is claimed that intelligence *has* evolved more than once – in birds, for example, and cetaceans.[9] According to that view, humans are just exceptional outliers in a continuum of intelligence, our amazing mental prowess the result of natural evolutionary amplification over millions of years. But this is contentious: humans are very biased in seeking human-like traits in other animals and anthropomorphizing their significance. Birds and cetaceans are certainly very clever in their own way, but the only intelligence that matters in the SETI game (as currently played) is the high-technology sort, because it's based on the principle of 'by their instruments ye shall know them'. There isn't a shred of evidence that, left to their own devices, birds or cetaceans would eventually write down Einstein's general theory of relativity or invent lasers.

The upshot of these arguments is that there is wide scope for disagreement. There *may* be a deep law of nature that drives living systems towards greater complexity, with big brains and intelligence being one consequence. But no such law is known to science, in spite of the widespread belief that it may exist. It is also possible that evolutionary convergence is so strong, and advanced intelligence has such good pervasive survival value, that it will sooner or later inevitably evolve, barring major calamities. However, in the absence of a second sample of life and a second evolutionary history to compare with ours, this is mere wishful thinking.

IS SCIENCE INEVITABLE?

Suppose we grant that high intelligence is in fact common in the universe. The next question of interest to SETI researchers is what proportion of those intelligent species proceeds to discover science, invent high technology, and engage in long-range communication. It is certainly fashionable, partly for reasons of political correctness, to assert that, here on Earth, *any* human society would be bound to discover science and technology in the fullness of time. To say otherwise seems to be implying the superiority of European civilization, where science as we know it began, and this is regarded by some people as racist and chauvinistic. Personally, I have always been sceptical of the claim that 'science is inevitable'. The problem is that science works so well, and is so much a part of everyday life, that people tend to take it for granted. The scientific method, taught (mostly badly) to every school student, comes across as a thoroughly obvious procedure: experiment, observation, theory – what could be a more natural way to find out how the world works?

The 'obvious' view of science is seen to rest on flimsy foundations when placed in a historical context, however. Science proper emerged in Renaissance Europe under the twin influences of Greek philosophy and monotheistic religion. The Greek philosophers taught that humans could come to understand the world by the exercise of reason, which achieved its most disciplined form in the rules of logic and the mathematical theorems that followed therefrom. They asserted that the world wasn't arbitrary or absurd, but rational and intelligible, even if confusing and complicated. However, Greek philosophy never spawned what today we would understand by the scientific method, in which nature is 'interrogated' via experiment and observation, because of the Greek philosophers' touching belief that the answers could all be deduced by pure reason alone. The Greeks' remarkable advances in reason and mathematics were nurtured for centuries during the European Dark Ages by Islamic scholars, without whom it is very doubtful that science and mathematics would have taken root in European culture in medieval times. An echo of the Islamic phase survives in modern terms like algebra and algorithm, and in the names of familiar stars such as Sirius and Betelgeuse. In spite of the importance of the Islamic phase in the lead-up to science, for some reason (possibly political or social) Arab scholars did not go on to formulate mathematical laws of motion or carry out laboratory experiments in the modern sense of the term.

Meanwhile, monotheism increasingly shaped the Western world view during the formative stages of science. Judaism represented a decisive break with almost all contemporary cultures by positing an unfolding cosmic narrative based on linear time. According to the Judaic account, the universe was created by God at a definite moment in the past, and developed in a unidirectional series (creation, fall, trials and tribulations, Armageddon, salvation, judgement, redemption . . .). In other words, Judaism has a cosmic story to tell, of a divine plan revealed through historical sequence. This was in sharp contrast to the prevailing view that the world is cyclic: the rotation of good times and bad times, the rise and fall of civilizations, the revolving wheel of fortune. Even today, the unidirectional linear-time world view of Western civilization rests uneasily with other cultural motifs, such as the dreaming of the Australian Aborigines or the cyclicity of Hindu and Buddhist cosmologies.[10]

The concept of linear time, and a universe created by a rational being and ordered according to a set of immutable laws, was adopted by both Christianity and Islam, and was the dominant influence in Europe at the time of Galileo. The early scientists, who were deeply religious, regarded their work as uncovering God's plan for the universe, as revealed through hidden mathematical relationships. What we now call the laws of physics they saw as thoughts in the mind of God. Without belief in a single omnipotent rational lawgiver, it is unlikely that anyone would have assumed that nature is intelligible in a systematic quantitative way, mirrored by eternal mathematical forms. The scientific method itself verged on being an occult practice at the time of Newton, and was conducted after the fashion of a secret society. Writing coded symbols on pieces of paper and subjecting matter to 'unnatural' experimentation in the sanctum of special laboratories is an arcane procedure by any standards. So science, though considered natural enough today, was little different from magic when it was first established.

Suppose an asteroid had hit Paris in 1300 and destroyed European culture. Would science *ever* have emerged on Earth? I have never heard a convincing argument that it would. It is often remarked that in medieval times the Chinese were technologically far more advanced than the Europeans, which is true. So why did the Chinese not go on to become true scientists? Part of the reason is that traditional Chinese culture was not steeped in the monotheistic notion of a transcendent lawmaker.[11] Outside the monotheistic world, nature was perceived as ruled by the complex interplay of competing influences in the form of gods, agents and concealed mystical tendencies. In medieval China, no clear distinction was drawn between moral laws and laws of nature. Human affairs were inextricably bound up with the cosmos, forming an indivisible unity. For the pagans of Europe and the Near East, who were in competition with Christianity and Islam at their formative stages, knowledge of the cosmos was to be gained through 'gnosis', a mystical communion with the creator, rather than through rational enquiry. Could gnosis eventually lead to science? I don't think so. Unless you *expect* there to be an intelligible order hidden in the processes of nature – fixed and analysable by mathematics – there would be no motivation to embark on the scientific enterprise in the first place.

Here we reach a key subtlety about the scientific method, which is the role that theory plays in physics. The power of theoretical physics stems from the recognition that there are deep interconnecting principles in nature. When Newton saw the falling apple, he didn't just see an apple fall; he perceived a set of equations linking the motion of the apple to the motion of the Moon. 'Theoretical physics' *does not* mean 'having conjectures about physics'. It means establishing an elaborate interlocking system of specific mathematical equations to capture aspects of physical reality that on casual inspection we would never guess are related, and then modelling those relationships quantitatively. No other science possesses this underpinning. There is no 'theoretical biology', let alone 'theoretical sociology' or 'theoretical psychology', in the physics sense of the word theory. There are ideas, conjectures, simple mathematical models, organizing principles, paradigms and so forth, but no true law-like mathematical *theory* (at least, not yet). The spectacular success of physical science derives from the fertile interplay of theory and experiment. Without minds prepared by the cultural antecedents of Greek philosophy and monotheism (or something similar) – and in particular the abstract notion of a system of hidden mathematical laws – science as we know it may never have emerged.

It is sometimes claimed that, even without a belief in a pervasive immutable law-like order in nature, any sufficiently long-lived society would stumble on science eventually, simply from trial and error. After all, the Chinese discovered the compass without a clue about how the Earth's internal dynamo generates a magnetic field or how that field interacts with electrons in the compass. Perhaps, then, the use of increasingly sophisticated tools would sooner or later lead to nuclear power and spacecraft and radio communication. For technology, it's enough to know *that*, without knowing *how*. Well, obviously it's possible in principle to discover, step by step, that certain causes produce certain effects. The true power of science, however, is that it leads us to *design* novel contraptions based on *understanding* the principles that govern them. With trial and error, one can perfect existing tools and devices, but without a sound theoretical basis, there is no reason to even go looking for most of the things that now dominate modern science. Why would one expect there to exist neutrinos or gravitational waves, for example, which almost all pass right through the Earth without having any measurable effect at all? Why look for dark matter or dark energy, which astronomers deduce from very careful observations using satellites and large telescopes, but which make sense only when suitably interpreted through layer upon layer of mathematical theory? Why build a particle accelerator unless you had reason to suspect that hitherto unknown and invisible particles like *W* and *Z* had a good chance of *being there*? Of course, there is a finite probability that a race of sentient beings without science may, by pure accident fuelled by curiosity, put together a radio telescope or a particle accelerator without the slightest idea of what they were doing or what the outcome would be, and have no actual understanding of what they found when they found it. Possible, yes, but the scenario is so ridiculous it cannot be taken seriously. It's as silly as saying that someone with no musical appreciation or ability will one day accidentally write a symphony.

I concede there may be some deep, as yet undiscovered, principle of social organization that says, roughly speaking, given a race of curious beings (and curiosity is certainly a general biological trait), then over time science is inevitable. It might be the case that human history has been channelled down the path of enlightenment and discovery by the unseen hand of such unknown laws of complexity and organization. (I shall have more to say about this conjecture in Chapter 8.) On the face of it, however, there seem to have been many contingent features – political, religious, economic and social – that went into the development of the modern scientific method. It could be that history is simply a series of random and unforeseeable accidents, one of them being the felicitous conjunction of Greek philosophy and monotheism in medieval Europe. If we do discover an alien civilization that found science, it would be strong evidence that there are indeed universal laws of social and intellectual organization, just as there are universal laws of physics. But without good reason to believe in such laws, the fashionable claim that 'science is inevitable' strikes me as totally without foundation.

THE DRAKE EQUATION

A good way of summarizing the discussion so far is to gather together the various factors that collectively determine the expected number of communicating civilizations existing elsewhere in our galaxy at this time. The result is known as 'the Drake equation', and was first written down by Frank in 1961 (see Fig. 5). It is not so much an equation in the conventional mathematical sense, more of a way to quantify our ignorance. I will ignore the usual rule of popular science writing that says no mathematics other than $E = mc^2$ are allowed under any circumstances, on the basis that the Drake equation isn't a real equation anyway. So here it is:

$$N = R^* f_p n_e f_l f_i f_c L$$

What do all these symbols mean? Let me give the definitions one by one:

R^* = rate of formation of sun-like stars in the galaxy
f_p = fraction of those stars with planets
n_e = average number of Earth-like planets in each planetary system
f_l = fraction of those planets on which life emerges
f_i = fraction of planets with life on which intelligence evolves
f_c = fraction of those planets on which technological civilization and the ability to communicate emerges
L = the average lifetime of a communicating civilization.

The number N on the left-hand side of the equation represents how many 'radio-active' civilizations are out there in the galaxy. Because traditional SETI focuses on radio signalling, what counts as a communicating civilization for the purpose of the Drake equation is simply one that possesses radio technology. There might be better ways to send signals across space, or there might be advanced alien civilizations that prefer not to engage in long-range communication, by radio or otherwise. But if there are, we won't spot them using radio telescopes.

The symbols on the right-hand side of Drake's equation are quantities we need in order to estimate – guesstimate would be more apt – the number N. Let me discuss each of them in turn.

Fig. 5. Frank Drake and his eponymous equation.

The first term, R^*, is the rate of sun-like stars being born per year in our galaxy. Why just in our galaxy? The reason is that receiving radio signals from beyond the Milky Way looks extremely unlikely, given the greater distances involved, although it is certainly not impossible. Anyway, let's go with the restriction for now. The accumulated *total* number of sun-like stars in the galaxy today is known quite well to astronomers (by simply pointing a telescope and counting, then scaling up using simple statistics). The answer is around 10 billion, depending a bit on just how 'sun-like' a star really needs to be to support life. But the number isn't fixed: stars are born and stars die, and so it has been since the Milky Way began forming about 13 billion years ago. For example, about seven new stars a year are currently being added to the Milky Way on average, though that number has changed somewhat over the course of galactic history.[12] The specifics don't matter. The point is that the uncertainty in the value of R^* is relatively small.

The next symbol, f_p, is the fraction of those stars that have planets. Back in 1960 when SETI began, this quantity was unclear because nobody could be sure how planets form. One theory suggested the solar system was made from material dragged off the sun by a passing star – surely a very rare occurrence, implying that f_p would be exceedingly small. Another theory supposed that the planets were made from matter concentrated in a disk or nebula of gas and dust swirling around the proto-sun. Drake – ever the optimist – went with the latter theory, and estimated $f_p = 0.5$, i.e. half of sun-like stars have planets. For decades there was little help from observation, but today astronomers are able to detect planets going around other stars, using techniques I discussed briefly in Chapter 1. The observations indicate that the nebula theory is right and that most stars have planets of some sort. In fact, Drake might have slightly underestimated the number.

Actually, the original Drake equation left out of account an entire class of planets, the importance of which has only recently been appreciated. Theoretical analysis of planetary motion suggests that orbits can be destabilized by planets 'ganging up', resulting in objects being flung out of a star system altogether. As a result, there could be many 'rogue planets' wandering the dark interstellar spaces, perhaps accompanied by a retinue of moons. Quite possibly our solar system started out with more than the eight (or nine) planets we see today, the rest being ejected. An enduring memory from my childhood is of the fortnightly BBC television fantasy *The Lost Planet*, screened in 1954. It featured a journey in an atomic-powered spaceship to the wandering planet of Hesikos, which temporarily entered the solar system from deep space. Hesikos turned out to be inhabited by telepathic humanoids. It was certainly a riveting story for an eight-year old, but the idea of a planet meandering 'lost' through the galaxy struck me at the time as the weakest link in the narrative. Lo and behold, it seems not to be so daft after all. Some astronomers estimate that there could be *billions* of rogue planets adrift in the Milky Way, so the Drake equation needs modification to take them into account.[13] Anyway, adding up both the tethered and loose planets suggests a tally of somewhere in the region of a trillion in our galaxy.

For life as we know it to arise, a planet has to be 'Earth-like'. The factor n_e in the Drake equation stands for the number of planets in a star system able to support life (i.e. 'Earth-like' planets – hence the subscript *e*). Drake initially picked 2 for the value of n_e, which is to say, an average of two Earth-like planets per planetary system. What do the observations show? In the case of the solar system, Earth and Mars would qualify. As far as extra-solar Earth-like planets are concerned, none has so far been discovered. But that should soon change when the results of the Kepler mission become available. Various more ambitious space-based planet-finding instruments are being planned, and it is possible we shall have acceptable images of other earths out to, say, fifty light years, within a decade or two. Almost certainly there *are* many Earth-like planets in the galaxy, but putting a precise number to it is hard. Somewhere between 1 and 10 per cent is my estimate of the fraction of planets in sun-like star systems that at least resemble Earth in their temperature, atmospheric pressure and surface gravity. That is lower than Drake's original figure, but not drastically so, and still amounts to billions of Earth-like planets.

Next comes the really hard part. The factor f_l is the number of Earth-like planets on which life arises. As I have been at pains to point out, that number is hugely uncertain. SETI enthusiasts such as Frank Drake and Carl Sagan put $f_l = 1$. In other words, they assumed that if a planet was like Earth, then life was bound to arise in due course – de Duve's cosmic imperative. On the other hand, sceptics like Jacques Monod chose f_l very close to zero. If we discover a shadow biosphere, we might be able to settle the matter in favour of a number close to 1. But for now, we are largely in the dark.

The factor f_l – being the fraction of planets with life on which intelligence evolves – I discussed earlier in this chapter. Sagan took the astonishingly optimistic value of 1, implying that intelligence is inevitable sooner or later, once life gets going. Originally, Drake assigned the slightly more conservative, yet still hopeful, figure of 0.01 for f_l. However, I have stressed that this number is also highly uncertain, as is f_c, the fraction of planets with intelligent life on which science and telecommunications develop.

HOW LONG DO TECHNOLOGICAL CIVILIZATIONS LAST?

The final factor in the Drake equation is the average lifetime of a communicating civilization. To appreciate the significance of this, imagine a town in which each homeowner switches his lights on and off for ten seconds, just once, at a time of night chosen randomly for each dwelling. Now ask how likely it is that two houses in the town will be lit up at the same time. If there are only a hundred houses in the town, chances are that no two houses will be illuminated together. Lights will come on and go off, randomly across the town, but probably never simultaneously. If the lights get left on for a minute rather than ten seconds, or if there are 10,000 houses in the town rather than a mere one hundred, then there is obviously a better chance of simultaneous illumination. Now think of communicating civilizations that way. They come and they go; they 'light up', then fade away. Right now, human civilization is 'lit up'. We'd like to know whether anyone else in the galaxy is going through their radio communication phase *now*.[14] It's no help when pursuing SETI radio searches to know that thousands of communicating civilizations may have come into existence in the Milky Way, but have long ago vanished, their transmissions ceased, or that thousands more will arise in the far future when humanity may have gone. The goal of traditional SETI is to acquire some cosmic company at *this* epoch. And the probability of success hinges on the term L in the Drake equation, the length of time an average alien civilization broadcasts radio signals. The bigger the value for L, the greater the chance that another civilization is on the air at this time.

Back in 1961, Drake picked L = 10,000 years. Sagan, who was depressed about human stupidity in relation to nuclear war and environmental damage, thought 10,000 years might be a bit optimistic. Michael Shermer of the Skeptics Society estimated that human civilizations are inherently unstable and typically collapse after only a few hundred years.[15] Some biologists have argued that the average lifetime of a mammalian species is a few million years, and this sets a quite general upper limit on the expected duration of our civilization. Of course, nobody really knows. Personally, I think all the arguments concerning L are naïve and irrelevant, especially the biological one. Darwinian evolution was already suspended with agriculture, and is now completely superseded with the advent of modern medicine, democratic rights, genetic engineering and biotechnology. Human civilization might yet succumb to a natural catastrophe, such as an asteroid impact or a species-jumping killer pandemic, or as a result of manmade disasters like nuclear war. But there is certainly no inevitability of such a thing, and if we make it through the next few centuries, we could be set fair for the indefinite future. I see no reason why, once an advanced extraterrestrial civilization is established, it shouldn't endure for an extraordinary length of time – millions or tens of millions of years or more. So this is one term in the Drake equation where I am more optimistic than the pundits.

Of greater relevance to traditional radio SETI is the question of whether the electromagnetic footprint of a civilization will also endure for an extraordinary length of time. Humanity has been broadcasting radio signals for about a century. Our most powerful emissions come from military radar. After that, it's TV stations. In the early days of SETI, scientists predicted a relentless rise in radio traffic, as wealth and technology advanced. But what happened was quite the reverse. First, point-to-point communications became dominated by low-power satellites directing their signals earthward. Second, the bulk of telecommunications shifted away from radio to buried optical fibres. If ET is monitoring our radio traffic, it will seem to have risen to a peak in the late twentieth century and then begun to fade. In another hundred years, there may be no substantial radio output from Earth. (Radar might still be used, plus the occasional command to a space probe.) So unless an alien community has a deliberate policy of transmitting powerful radio signals, it is entirely possible that the galaxy is bustling with advanced civilizations yet has no detectable artificial radio signature. It has been estimated that if we built a radio telescope 100 kilometres (60 miles) in diameter, it would be so sensitive we could detect a TV station as far away as Sirius, so it wouldn't matter whether ET were beaming messages directly at us or not. But if Sirius TV is delivered via cable, we'd be out of luck. Eavesdropping on an extraterrestrial civilization on the premise that the aliens may still be using 1980s human technology is a hard sell. (I shall return to this topic in Chapter 5.)

Anyway, for what it's worth, if Drake's figure $L = 10{,}000$ is adopted, together with his estimates for all the other factors in his eponymous equation, one obtains the bottom line result $N = 10{,}000$; that is, there should be 10,000 civilizations in the galaxy at this time capable of communicating with each other (and us) using radio technology. Which seems very exciting. Ten thousand extraterrestrial civilizations on the air right now! If we knew that for a fact, SETI would be an urgent priority. 'Let's find them!' everyone would say. But as I have explained, although many terms in Frank's equation are known rather well, and one at least (*L*) was in my view seriously underestimated, the equation is utterly dominated by two factors about which we know almost nothing – f_l, the fraction of Earth-like planets on which life emerges, and f_i, the fraction of those on which intelligence evolves. In my view the former is much more problematic than the latter. If life gets going, intelligence is at least in with a chance. It could be that intelligence is, after all, wing-like rather than trunk-like; it's not too incredible. But it's entirely possible that life's origin is so freakish it has happened only once, and we are it. At this time we have no scientific grounds for refuting this position. There is to date not a shred of evidence that 'nature favours life', that there is a 'life principle' directing murky chemical soups towards the grandeur of biology. And since we haven't a clue about *how* life actually emerged, then unless and until we find either a shadow biosphere or strong evidence for life on an extra-solar planet, we can't even bracket f_l by concocting optimistic and pessimistic numerical estimates. At this stage of the game, the fraction could be anything at all between 0 and 1.

THE PERILS OF USING STATISTICS OF ONE

Given that our galaxy contains about 400 billion stars, a plausible guess at the number of Earth-like planets around sun-like stars might be a billion. If Monod is right, only one of these planets possesses life. If de Duve is right, most of them do. What about a middle position? Might our galaxy contain, say, a million planets with life?

There is a persuasive argument against the middle position. The 'other earths' don't just sit there for eternity waiting for biology to happen; there is a finite window of opportunity for life to emerge. Life as we know it requires a stable star like the sun to provide energy and maintain habitable conditions on a planet. But stars can't shine for ever;

sooner or later they run out of fuel and die. At 4.5 billion years of age, our sun is about half-way through its complete life cycle, having already consumed a large fraction of its nuclear fuel. In another billion years or so it will begin to feel the effects of fuel starvation, as a result of which it will swell up and slowly incinerate our home. (In astro-speak, it will start turning into a red giant star, a phase that presages death by collapse into a white dwarf.) A similar story is played out by stars throughout the galaxy. So if life is to emerge on a planet orbiting a given sun-like star, it has to do so in the 5- to 10-billion-year time window bracketed by the formation of the star and burn-out. Assuming that biogenesis occurs randomly on habitable planets, there will be statistical scatter, or a range of values for the amount of time needed to make it happen. But let's focus on the *average* time. If the average time is short – if life is quick and easy to form – there will be plenty of opportunity for it to begin on many planets (de Duve's view). On the other hand, if the expected time for biogenesis is much greater than 10 billion years, life may never get started at all on a given Earth-like planet. If it did, it would be against the odds – a lucky fluke. Expressed more scientifically, it would be a very rare fluctuation, an outlier in the statistical spread. In that case it is entirely possible that it happened on only a single planet in the galaxy, which would be Earth (Monod's view).

Turning now to the intermediate case of life arising on (say) a million planets in a galaxy like ours, the expected time for biogenesis to occur would have to be neither much shorter nor much longer than the average habitability window of a planet – say between one tenth and ten times. Is this reasonable? Let's consider what it entails. The length of the habitability window, which is bracketed by the duration that a star burns in a stable manner (call it $T1$), hinges on a variety of factors, such as the rate of nuclear reactions in the star's core, the efficiency with which heat is transported to its surface and the overall mass of the star. Now consider how long it might take for life to arise on an Earth-like planet (call that $T2$). For the moment I am considering only simple microbial life, not intelligent life. Of course we don't know the number $T2$, but if the middle position of a million planets with life is correct, then the time needed for biogenesis to occur would have to be a few billion years (i.e. comparable to $T1$, the lifetime of the stable phase of an average star): life would then fail to start in time on some

Earth-like planets, on many it would form near the middle of the window of opportunity, while on a few it would begin just before the planet became uninhabitable. Such a scenario, although certainly possible, would, however, represent a very improbable coincidence. The time required for life to emerge from non-life has, on the face of it, nothing to do with the factors that determine the lifetime of a star, such as the rate of nuclear reactions. As far as we can see, life is a product of physical processes – involving atomic and molecular physics, chemistry and geology – that are altogether different from those taking place inside stars. So why should the durations $T1$ and $T2$ possess the roughly *equal* values that would be required for a million planets to generate life, when the two timescales have no causal connection? There is no obvious reason why one number isn't much larger than the other. It could of course be that $T1$ and $T2$ are comparable in value merely by chance; coincidences are allowed in science, but they should be the explanation of last resort.[16] If coincidences are rejected, then the conclusion must be that the expected duration of time for life to emerge is quite likely to be very much less than the lifetime of a star, or very much more.

Which is it to be? All we have to go on is life on Earth – a sample of one. Drawing statistical conclusions is therefore risky, but that hasn't stopped people from trying. It was pointed out by Carl Sagan that life began on Earth rather quickly: 'the origin of life must be a highly probable affair; as soon as conditions permit, up it pops!' he wrote.[17] Sagan was referring to the fact that Earth suffered severe bombardment until about 3.8 billion years ago, and according to the fossil record life had become firmly established within 300 million years (see Fig. 6). That suggested to Sagan that whatever unknown process produced life, it was fast, and therefore life might be expected to arise with comparable rapidity on other Earth-like planets.

Sagan may be right, but unfortunately there is a serious complication. The reason that life on Earth is chosen for our single statistical sample is precisely because we ourselves are a product of it. Earth harbours not merely life, but *intelligent* life, at any rate intelligent enough to concoct arguments about biogenesis. To attain that level of intelligence, life has to evolve to a high level of complexity, and it must do this within the few-billion-year habitability window during which

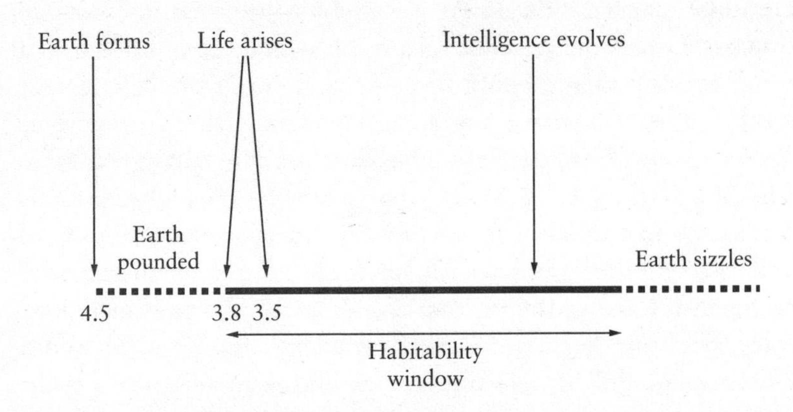

Fig. 6. Life established itself on Earth rapidly once conditions became suitable. However, had it not done so, humans might not have evolved before the habitability window closes in about 800 million years' time. Numbers express billions of years before the present.

the sun burns stably. Crucial steps on that road included the emergence of multicellular organisms (which took over two billion years), the evolution of sex, the formation of nervous systems and the development of large brains. In between were myriad smaller steps, some hard, others easy. Obviously, unless all the steps had been completed within a few billion years, humans (or animals of comparable intelligence) would never have evolved enough complexity to deliberate on scientific matters. In other words, life on Earth *had* to get going pretty fast, or there wouldn't have been enough time for intelligent observers like us to hit the scene before the sun became a red giant. So life's prompt appearance on Earth may not after all be indicative of the general situation; it could have been a highly atypical set of events which has been selected for observation and scrutiny by the very observers it created.

THE GREAT FILTER

The rough-and-ready argument I just outlined was placed on a sound mathematical footing in 1980 by the British cosmologist Brandon Carter,[18] and subsequently refined by the economist Robin Hanson.[19]

Carter and Hanson imagined a large ensemble of 'experiments' in which nature has a chance to produce intelligent life, and they noted that if the expected time for intelligence to evolve were much *less* than the lifetime of a typical star (say, a mere one million years), it would be hard to see why it has taken billions of years for it to run its course on Earth. One would have to make a case that although intelligent life is common in the universe, for some peculiar reason the evolution of intelligence on Earth was atypically delayed. On the other hand, suppose the expected time for the evolution of intelligence is much *longer* than the lifetime of a typical star, yet in spite of the highly adverse odds intelligence *does* in fact evolve (as it did on Earth), then the time it would take to complete this highly improbable process would most likely be close to the total permitted duration, i.e. the length of the habitability window. And that is indeed what we observe: the evolution of intelligent life on Earth has 'used up' about 4 billion years of the roughly 5-billion-year window of opportunity, before Earth gets fried by the swelling sun (see Fig. 6).

Carter and Hanson were able to quantify this idea precisely. Here is the gist of their result, which follows in a straightforward manner from the equations of probability theory, but the curious reader will have to consult the original papers for the actual proof. Assume that several vital steps take place on the road to intelligence, and that each step is so improbable it would take, on its own, far longer on average than the lifetime of a typical star.[20] Hanson calls this obstacle race for life 'The Great Filter'. Suppose there are *N* such steps, and that, *against the odds*, intelligent life *does in fact arise*. Then the equations show that the expected time between each highly unlikely step is about 1/Nth of the habitability window, with another 1/*N*th left before the window closes. I have depicted this result in Fig. 7. Curiously, the gaps between the steps are independent of just *how* hard the steps might be, so long as they are all very hard. (Intuition might suggest that if step *A* had a one in a million chance and step *B* a one in a billion chance, then, in the event that both these steps actually did happen, *A* would happen about a thousand times faster than *B*. But not so.)

What can we say about the number *N* if we apply the Carter–Hanson argument to the actual situation on Earth? If our understanding about the sun's evolution is correct, then (according to the best estimates)

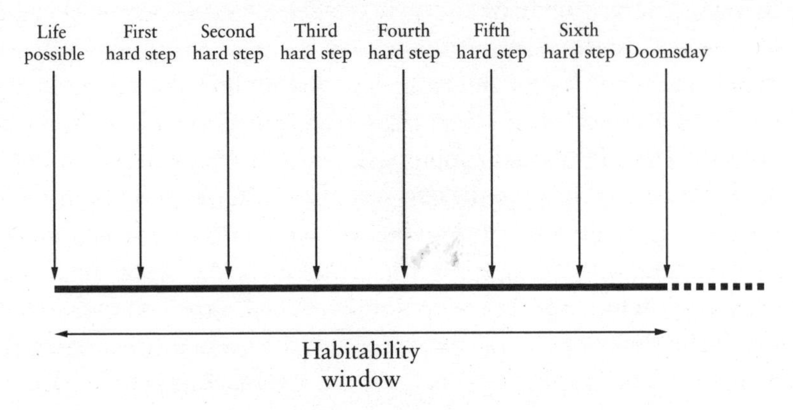

Fig. 7. The Great Filter, in the case that there are six extremely improbable steps on the road to intelligent life, and assuming that intelligence nevertheless emerges, against the highly adverse odds, before the multi-billion-year habitability window closes. The key result, proved using probability theory, is that the durations between gaps are (roughly) equal, and of the same duration as the time left before doomsday, when the habitability window closes. Knowing how long we have got on Earth before doomsday serves to fix the size of the gaps, and hence the number of steps. Using 800 million years for the time left yields six steps, as shown here. Plausible unlikely biological transitions can be found for each step. The data fit better if the first step occurs on Mars and life is subsequently transferred to Earth.

there's about 800 million years to go before our planet is too hot to support intelligent life. That suggests *N* is about 6 (this being the total duration of the window – 5 billion years – divided by the expected time left – 800 million years). That is to say, there were about six crucial but highly improbable hurdles to surmount en route to intelligent life, each of which should have taken place roughly 800 million years apart. How does that compare with the fossil record? Quite well, in fact. Major unlikely steps can be identified with, first, the origin of life itself; second, the evolution of photosynthesis in bacteria 3.5 billion years ago; third, the emergence of 'eukaryotes' (large, complex cells with nuclei) about 2.5 billion years ago; fourth, sexual reproduction about 1.2 billion years ago; fifth, the explosion of large multicellular organisms 600 million years ago; and, finally, the arrival of brainy hominids in the recent past. This all looks good, except for the first hurdle. Even allowing for the crude approximation, it seems to be a serious mismatch, because life took nowhere near 800 million years to get started on Earth. Rather, it had already emerged only 200–300 million years after the end of the cosmic bombardment – which was Sagan's point about life 'popping up' with almost indecent haste. So does this awkward fact demolish Carter's argument? Not quite. Carter has countered that we cannot be sure life actually began on Earth; it might have started on Mars and come to Earth inside ejected Martian rocks, establishing its first toehold on our planet only when the bombardment dwindled. If he is right, then the window of opportunity for life to arise could be extended back from 3.8 to 4 billion years ago or even earlier, because Mars was ready for life sooner. All the steps in the Great Filter, including the first, would then be spaced out by roughly the predicted 800 million years.[21]

Earlier I discussed how the intelligence hurdle wasn't surmounted readily on Earth – it took over 200 million years of brain evolution among land animals before hominids evolved. That was bad enough. But Carter's reasoning suggests a far more pessimistic conclusion. The predicate of his argument, remember, is that the average, or expected, time for intelligent life to arise is much longer even than the several-billion-year habitability window offered by a typical star like the sun. So the fact that intelligence took over 200 million years to evolve on Earth, slow though that may seem to us, should be regarded (according to Carter) as a fluke, a statistical outlier, an event lucky to have happened at all in so short a window. And the upshot of this 'lucky Earth' conclusion is that the vast majority of other sun-like stars will not share our system's good fortune. They will fail to possess planets with intelligent life. If Carter is right, then, Earth is a *very* rare exception, and the emergence of intelligent beings like humans is a freak event, just as Monod maintained.[22]

Though Carter's argument seems to knock the stuffing out of SETI, many of my colleagues are suspicious of the underlying reasoning. A popular objection is that we can't use guesses about the future (for example, how long before Earth becomes a roasted crisp) to reason about the past. In fact, this is a spurious objection: probability arguments are perfectly valid applied to both past and future events so long as all other factors remain unchanged through time. But suppose all other factors do not remain unchanged. For example, what if galactic-wide cosmic catastrophes frustrate the appearance of intelligent life for billions of years, and then abate? One of the most violent events in the universe is a gamma ray burst. These unpleasant cataclysms are probably caused when massive stars implode to form black holes, releasing a huge spray of energy in the form of electrically charged particles directed along pairs of oppositely oriented narrow beams. The charged particles in turn generate intense gamma radiation (high-energy photons), that paint the galaxy in arcs, like cosmic death rays, as the black holes rotate. If one of the gamma ray beams sweeps over a planet, it could annihilate all complex surface life. Gamma ray bursts are observed using a satellite named Swift, which registers hundreds of events per year. They would have been more common in the past, and could conceivably have prevented intelligent life from evolving anywhere in the galaxy for some billions of years. If so, then maybe under ideal conditions (i.e. not menaced by gamma rays) intelligence isn't all that improbable after all. The fact that it took a long time to evolve on Earth would have a ready physical explanation (Earth was zapped by gamma rays), and Carter's conclusion that intelligence is highly unlikely even after *tens of billions* of years would be weakened. So the jury is still out on just how serious Carter's line of reasoning might ultimately turn out to be, once we understand all the factors that go into determining what it takes for intelligent life to arise.

ARE WE DOOMED?

Before moving on from the battle of the probabilities, there is a final twist that needs to be considered. If the eerie silence is taken as *prima facie* evidence that we are alone (in the sense that we are the only intelligent beings in the universe) then it could be that the steps leading up to intelligent life are so unlikely they have happened only once.[23] But there is a second possible explanation for the silence, one that I mentioned in the previous chapter. Perhaps intelligent life and technological civilizations are inherently unstable, and so do not survive for long enough to make contact with each other. If *that* is the correct explanation, then it is bad news for humanity. It implies that, if Earth is typical, we can expect to go the same way as the aliens, following our cosmic cousins into oblivion fairly soon – or at least, before we get to broadcast to the galaxy. And of course it's not hard to identify potentially calamitous hazards that could wipe us all out – nuclear war, killer pandemics, comet impacts, social and economic disintegration . . .[24]

How can we determine which of the two explanations for the eerie silence is the more likely: lucky Earth, or doom soon? In the absence of any evidence either way, both scenarios are equally plausible. But that state of ignorance could soon change. If the silence is real, and not just the result of bad luck or poor search strategy, then something acts to filter out most advanced technological civilizations, either by preventing their formation in the first place or by annihilating them soon after they become established. In the former case the Great Filter lies in our past, and we lucky humans have evidently passed through that part of the filter. In the latter case, the filter lies in our future, which is ominous: we may not be so lucky going forward, and might well get 'filtered out'. Suppose we uncover evidence for life beyond Earth, from the discovery of microbes elsewhere in the solar system, for example, or from oxygen in the atmosphere of an extra-solar planet. It would then follow that the first step on the path to intelligence and technological civilization – the genesis of life from non-life – is not in fact a huge and improbable leap. We could then conclude that the Great Filter must lie *ahead* of the first step, a conclusion that would serve to tip the balance towards it lying in the future of the emergence of intelligence, and thus shortening the odds for an impending human apocalypse. The situation becomes even bleaker if we discover not just primitive life, but more complex forms of life beyond Earth, because additional steps on the path to intelligence would then be revealed as likely, rather than unlikely. It would have the effect of further weakening the case for the Great Filter lying in the past of intelligent life, and strengthening the likelihood of a dangerous future for intelligence. In short, if life *is* a cosmic imperative, then the great silence is indeed eerie; in fact, it is positively sinister as far as the fate of humanity is concerned. If ET isn't out there, we had better hope that *no* life is out there. Nick Bostrom, an Oxford University philosopher, sums it up bluntly: 'It would be good news if we find Mars to be completely sterile. Dead rocks and lifeless sands would lift my spirits . . . It promises a potentially great future for humanity.'[25]

In 1979 I was asked to write a script for the actor Dudley Moore, who played the role of a bewildered student in a BBC documentary called *It's About Time*. The narrative began with the famous paradox of the Greek philosopher Zeno, according to which an arrow could never reach a retreating target, for the following reason. No sooner will the arrow arrive at the place the target occupied when the arrow was unleashed than the target will have moved on a bit. And when the arrow reaches that new position, the target will have moved on again, and so on, ad infinitum. The TV version showed Dudley Moore running from the bowman, and then falling flat as the arrow struck him in the back, at which point the narrator commented wryly, 'So much for philosophy.' The philosophical arguments I have presented in this chapter, intriguing though they may be, are no substitute for hard data. They build grandiose cosmic conclusions from the slenderest of facts, and are only as good as the assumptions on which they are based. So long as there is no concrete scientific evidence for life beyond Earth, they are about all we can do. But SETI is fundamentally an experimental and observational programme, not an exercise in philosophy and statistics. A single discovery, like a single bow shot, could instantly overturn centuries of philosophical presupposition. An eerie silence is no reason to abandon the search for extraterrestrial intelligence. Rather, it provides a compelling reason to widen it.

5

New SETI: Widening the Search

Vision is the art of seeing what is invisible to others.

Jonathan Swift

THEY DON'T KNOW WE ARE HERE

The traditional approach to SETI is based on the belief that alien civilizations are targeting Earth with narrow-band radio messages. But in my opinion, this 'central dogma' simply isn't credible. The reason concerns the finite speed of light, and the fact that no signal or physical effect can propagate any faster. This absolute speed limit is a fundamental law of physics having to do with the nature of space and time. Unless our understanding of basic physics is badly wrong (in which case, much of the discussion about SETI is moot), we have to live with the restriction. To appreciate the implications, consider an alien civilization situated 1,000 light years away – close even by the standards of SETI optimists – and suppose that the aliens possess technology so powerful that they can observe the Earth in detail. What will they see? Well, they won't see us. They won't see our radio telescopes or our particle accelerators or roads or rockets. What they will see is Earth *c.* AD 1010. That date is well before the Industrial Revolution, at a time when the pinnacle of human technology was the clockwork. The aliens might see the Egyptian pyramids and the Great Wall of China. They would notice cities and signs of agriculture, but that is a far cry from interstellar telecommunication technology. The fact that humans had developed the use of building and agriculture might be promising, but it would certainly not guarantee the appearance of radio telescopes 1,000 years later (as opposed to, say, 5,000 or 50,000 years later). Therefore, there would be no reason for the aliens to begin transmitting radio signals our way in AD 1010. Better for them to wait until they know we actually have the means to receive the signals before going to the trouble of sending them.

How, then, will the aliens know when we are ready for their message? Well, when *our* first radio signals reach *them*. Human radio technology is about a century old. In about another 900 years those first weak signals will reach this imaginary nearby civilization, and if the aliens were continuously monitoring us with *very* sensitive equipment, and were quick off the mark, we might get their first message just before the start of the fifth millennium. There is no getting around the delay. In 'their' universe (that is, from the aliens' delayed-time perspective) human radio astronomers simply do not yet exist. Unless they can see into the future, there *is* no target technological civilization on Earth for them to signal at, and there won't be one for another 900 years. And if the alien civilization is even farther away – 10,000 light years, say – then the wait is that much longer. The upshot is that traditional SETI – probing the skies with radio telescopes looking for a message from the aliens – may well be a good idea, but we are doing it a few millennia too soon. The only let-out is if an alien presence is located much closer – within fifty light years. That would be amazing, but who knows? However, SETI astronomers have looked at every candidate star system out to that distance, and drawn a blank.

The foregoing conclusion, while depressing, isn't an argument against a broader strategy for SETI; it merely points up the futility of searching for messages that are deliberately directed at human civilization from a faraway source. A radio search of the sky might conceivably stumble across alien radio messages intended for someone else who happened to be located along our line of sight, a message coincidentally transmitted a long time ago that is traversing our astronomical neighbourhood at this time. Obviously that is a distant hope. Another remote possibility is that there are alien civilizations broadcasting messages indiscriminately and continuously to the entire galaxy – the galactic equivalent of the BBC World Service. But that would demand a stupendously powerful transmitter, and a level of determination and altruism we have no right to expect.

Another long-shot idea being touted by SETI researchers is the possibility of eavesdropping on routine domestic radio traffic leaking from another planet. Our own radio and TV stations broadcast at much lower frequencies than SETI searches – typically in the range 50–400 MHz. (SETI focuses on a wide-frequency band, but in the 1–2 GHz range.) However, a new class of radio instruments is being built that will cover the MHz range nicely, and with unprecedented sensitivity. Nearing completion in Europe is a system called LOFAR, for Low Frequency Array. It consists of 25,000 metal rod antennas located in several countries, linked together electronically so the data can be digitally amalgamated. Rather than hopping from source to source, LOFAR has the ability to watch large patches of the sky for months at a time, thus increasing the chances of spotting a continuous weak signal. The primary purpose of LOFAR is to study the end of the so-called cosmological Dark Age – the period immediately prior to the formation of the first stars. Because the universe has expanded greatly since that epoch (which was about 13 billion years ago), the wavelength of electromagnetic emissions has been stretched, so that at the receiving end (Earth) many interesting sources will have frequencies shifted down to the MHz range. LOFAR is not the only game in town. A more ambitious system with a similar concept and purpose, called the Square Kilometre Array (SKA), is slated to be built either in radio-quiet Western Australia or south-west Africa. As the name implies, this collection of antennas would cover an area totalling a square kilometre. While these highly sensitive instruments are going about their routine astronomical business, SETI researchers can piggy-back on them without disturbing their primary purpose.

Welcome though this new generation of instruments may be for SETI, it seems that neither LOFAR nor SKA is up to the alien-eavesdropping job, unless we get very lucky. In spite of their immense size, these instruments couldn't detect an Earth-strength television station even if it was located on a planet going around the nearest star. But there is a glimmer of hope. Abraham Loeb of Harvard University has estimated that a terrestrial-strength TV transmitter *could* be detected by the SKA up to several light years away if observations were accumulated continuously over a month, and assuming a way can be found to filter out terrestrial interference in the same waveband.[1]

Although that distance range encompasses many stars, it is still within our local neighbourhood, astronomically speaking. There is no hope of picking up a TV station at a distance of, say, 1,000 light years, unless its transmissions are much more powerful than their terrestrial counterparts.[2] A bigger problem awaits here too, one that I have already mentioned in Chapter 4. High-powered radio emissions are likely to be just a fleeting craze among emerging civilizations, if human experience is a guide. Already most of our TV channels are delivered by optical fibres. It is entirely possible that within a few decades Earth will be almost completely radio-silent, and our telecommunications will suffer almost no leakage into space. But a very old alien civilization might conceivably have its own reasons for continuing with domestic radio broadcasts, so it still makes sense for SETI to use LOFAR and SKA to search.

BEYOND THE PHOTON

Radio and laser signals are both electromagnetic – they use photons to convey messages. In principle, however, anything that goes from *A* to *B* could be used to encode a signal, so a broader SETI strategy should consider that alien signals might be transmitted in some other way. A technical problem faced by any means of signalling is that, if *A* and *B* are many light years apart, there might be obscuring material in the way, such as gas and dust. That is especially true in the plane of the galaxy, where dust is conspicuous in the form of dark lanes streaking across the Milky Way. Radio and laser light both have the advantage that, at certain wavelengths, this material is relatively transparent to them. Nevertheless, something with a greater penetrating power than photons might work better for interstellar messaging. One possibility is neutrinos, famous for their extraordinary ability to pass through matter. The snag is they tend to pass right through receivers, too. If ET is using neutrino beams to send messages, we have our work cut out to spot them.

For many years neutrinos remained purely theoretical, because there was no equipment sensitive enough to register them. That changed in the 1950s when intense neutrino fluxes emanating from nuclear reactors were finally detected. Although their interaction with matter is extremely weak, a neutrino will occasionally hit a nucleus and bring about a detectable transmutation. But the probability is exceedingly small: trillions of neutrinos sweep by for every one that registers a hit. Today, neutrino physics is very advanced. For example, neutrino beams are made at particle accelerator laboratories and shot through the Earth, to be picked up by instruments thousands of miles away. Huge detectors are being built consisting of kilometre-wide volumes of ultra-pure water (or ice), from which tiny flashes of light are emitted when neutrinos strike nuclei and create high-speed charged particles. The flashes are then amplified and registered by sensitive equipment. Physicists are constructing detectors in Antarctica, beneath the Mediterranean Sea and in Siberia's Lake Baikal, to explore the universe through 'neutrino eyes'. Bursts of high-energy neutrinos are expected from supernovae, black holes and possibly dark matter processes.

So in spite of the difficulty, humans do possess detectors that could in principle pick up an alien message encoded in a neutrino beam.

Neutrino signalling has been studied by Anthony Zee of the Kavli Institute of Theoretical Physics at the University of California Santa Barbara, and his colleagues,[3] who suggest that the aliens would opt for neutrino energies far above those generated naturally by the sun and stars. Because there are very few energetic neutrinos coming from any specific direction of space, a beam of high-energy neutrinos that passed our way would be highly conspicuous. Contrast this with energetic radio waves, which are generated by many compact astronomical sources; using radio, ET is in competition with the entire cosmos. Zee thinks the aliens could use a particle accelerator to collide and annihilate electrons and their antiparticles (positrons) to make a narrow beam of neutrinos that can be aimed at will. This is a tried and tested technique employed by terrestrial physicists, but the aliens need to do it at a much higher energy, a bonus being that the greater the energy, the easier neutrinos are to detect. Best of all would be an energy at which the transmitted neutrinos react particularly strongly with atomic nuclei, creating a spray of particles known to physicists as *W* bosons. (For the technically minded, this energy is 6.3 PeV.) If we saw *W* bosons being made that way, we would certainly take notice. To encode a message, all ET needs to do is use a type of Morse code. Admittedly the data transfer rate would be pretty pathetic, but as I shall now argue, that may not be so important.

BEACONS

Everybody is familiar with the computer, but few people know who invented it. Amazingly, the basic design of the universal computing machine was worked out as long ago as the middle of the nineteenth century by an eccentric English genius named Charles Babbage. Sadly his mechanical calculating engine, or Analytical Engine, was never completed. However, a replica of its precursor, the so-called Difference Engine, was made and operated by the Science Museum in London in time for Babbage's bicentenary in 1991.

Among Babbage's many other inventions and accomplishments is the now familiar signalling system for lighthouses. The principle is simplicity itself: a beam of light sweeps around in a horizontal plane and from a fixed point is seen to flash once or twice on each transit. The signal is not directed at anyone in particular, but whoever is sailing within sight of the lighthouse will notice it. The signal stands for 'Danger: navigate with care' and also 'Somebody is here.' That's about it: low total information content, but of enormous significance, at least for mariners.[4] Could an advanced alien civilization have constructed a similar beacon to sweep the galaxy?

Historically, the idea of signalling between planets using beacons predated radio SETI by at least a century. In 1802 the mathematical genius Karl Friedrich Gauss suggested creating huge shapes in the Siberian forest to attract the Martians' attention and signal our intelligence. His idea was to clear the forest and plant the interior with wheat, to form a pattern that signifies Pythagoras' famous theorem of geometry. Later, Percival Lowell dreamed up something similar, using oil-filled channels in the Sahara, which could be ignited at night. A variant on the 'big-geometry' theme was the proposal by the inventor and telescope maker Robert Wood, who wrote to the *New York Times* proposing an enormous black spot made from strips of cloth, which could be rolled up and unrolled periodically, making the spot appear to wink at our Martian neighbours! These early proposals all lacked the amplification and range to work beyond the confines of a single planetary system. But with the development of high-power radio and lasers, the way lay open to make a beacon that could signal across not just interplanetary but interstellar space.[5]

The possibility that alien civilizations might long ago have created powerful radio beacons, and that humans have the means to detect them, has been studied in detail by Greg and Jim Benford, twin physicists working in California. Greg is an astrophysicist and also an award-winning science fiction writer, while Jim is an expert on high-intensity microwave beam technology. The way the Benfords see it, ancient civilizations could have many reasons to build a beacon; for example, it could be a high-tech monument of pride to what may be a glorious but now long-vanished civilization. A beacon is also a great way to attract attention and simply make first contact: anyone detecting it would redouble their efforts at SETI. It could conceivably be an artistic, cultural or religious symbol, or even the cosmic equivalent of graffiti. It might be a cry for help, or, as with the humble lighthouse, a warning.

The Benfords have worked out the power requirements for microwave (rather than optical) beacons that operate by emitting intense, short-duration pulses – pings, if you like. Obviously it requires a lot less power to transmit a sporadic ping than a continuous stream of messages. While pulses are moderately harder to detect, they are considerably easier to transmit (although a beacon with galactic reach is still well beyond human technology). The starting assumption of the Benfords' calculation is that the cost per ping is something determined by fundamental physics, to which the alien builders are just as constrained as we are; presumably even a super-civilization wouldn't deliberately squander resources.[6] The Benfords have therefore analysed the problem 'from the point of view of the guys paying the bill', as they put it, and came up with what they think the characteristics of a beacon pulse would be, taking into account the capital costs of building the antenna and the operating costs of running it.[7] Efficiency favours higher frequencies, so they suggest 10 GHz is optimal; go above this and the background radio noise of the galaxy interferes. Most SETI observations have so far concentrated on a much lower band – around 1 or 2 GHz. There is a trade-off between the duration of each ping and the revisiting time between pings. A good compromise would be a burst of about one second's duration about once a year.

In contrast to the classic SETI target – a continuous narrow-band signal at a specific frequency – a beacon would show up spread across a range of frequencies in the form of a short blip, or perhaps a more attention-grabbing blip-blip. As it happens, many blips have been recorded throughout the lifetime of SETI, but very little follow-up has resulted, and for good reason. As we've seen in Chapter 1, the procedure when a radio telescope picks up something odd is to move the antenna off target, to make sure the signal fades (thus eliminating equipment malfunction), and then move it back on target again. If the signal is still there the second time, a partner radio telescope, preferably far away, is brought into play to confirm that the source is in fact astronomical (and not a local mobile phone, for example). All this assumes that the mystery signal will continue for long enough for the checking procedure to be completed, which in practice could take several hours. But if a telescope detects a momentary blip – there one moment, gone the next – the checking procedure isn't possible.[8]

A famous mystery pulse is the aptly named 'Wow!' signal, detected on 15 August 1977 by Jerry Ehman using Ohio State University's Big Ear radio telescope. The signal lasted for seventy-two seconds (rather a long pulse), and has not been detected again. Ehman discovered it whilst perusing the antenna's computer printout, and was so excited he wrote 'Wow!' in the margin (see Fig. 8). The signal has never been satisfactorily accounted for as either a manmade or a natural phenomenon. Another much-discussed transient event is an intense half-a-millisecond blip known as Lorimer's pulse, detected near the Small Magellanic Cloud by the Parkes radio telescope in Australia (see Plate 12). It was found by David Narkevic, an undergraduate student working for David Lorimer of the University of West Virginia. Lorimer wasn't looking for ET, but rather for astronomical objects called pulsars. The enigmatic pulse was discovered long after it was received, buried in data recorded from a routine search. Nothing similar has been observed again from that part of the sky. There is no consensus about the source, although it does appear to have come from a very long way away, far beyond the confines of our galaxy.

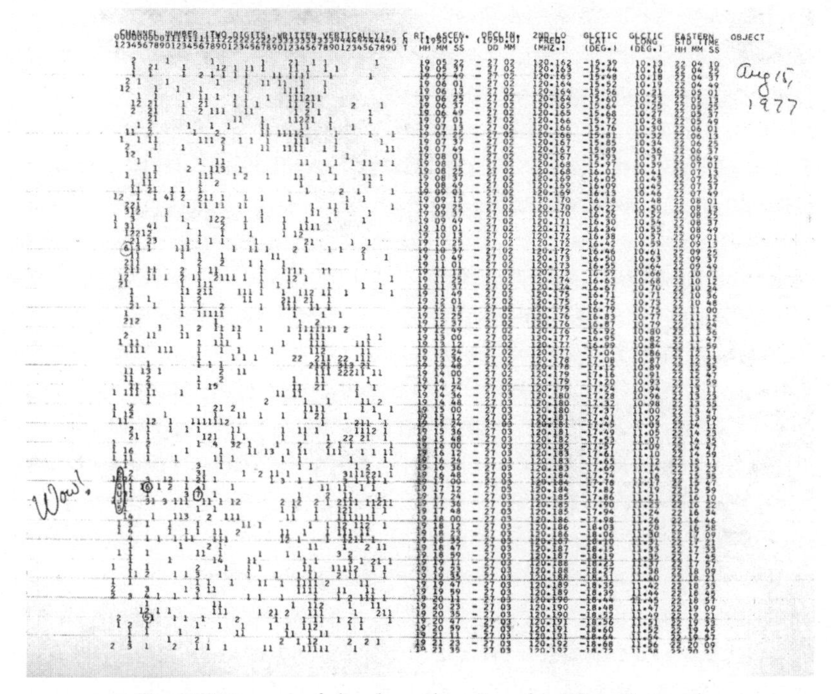

Fig. 8. Printout of the data showing the 'Wow!' signal.

The best guess is that it was caused by a violent black hole event of some sort.

Another possible source of radio pulses is exploding black holes. In 1975, Stephen Hawking concluded that black holes are not actually black, but radiate heat and, as a result of the energy loss, shrink in size, eventually evaporating away completely. Because the temperature of the black hole rises as the object shrinks, the evaporation is a runaway process, culminating in a frenetic final burst of high-energy particles, including many that are electrically charged. If this terminating explosion takes place in an ambient magnetic field, such as that of the galaxy, the charged particles will create a short but powerful electromagnetic pulse.[9] Direct searches for black-hole explosions using radio telescopes have yielded nothing so far.

The challenge for SETI is to discriminate between an artificial pulse and a natural one. If an alien civilization wanted to use pulses to attract attention, it would need to tag them with a signature of intelligence, such as a simultaneous transmission centred on several radio channels at frequencies that bear a noticeable arithmetic pattern. Existing SETI systems are not well adapted to dealing with such signals, because both the hardware and data analysis are mainly designed for continuous narrow-band sources. But there is no fundamental obstacle to conducting a search for pulses; the issue boils down to resources. Looking for transient events requires monitoring a slice of sky continuously for some time – say one year – because even if we can make an intelligent guess *where* in the sky the beacon might be located we don't know *when* it will next bleep. A pilot search for millisecond pulses is currently under way at the Allen Telescope Array, using a system called Fly's Eye, operated by the University of California at Berkeley. In the configuration employed, each of the forty-two currently operational dishes is pointed at a different patch of sky, giving very wide coverage in total. Unfortunately, as the aperture of the dishes is only 6 metres, the sensitivity is severely limited. Another dedicated search, known as Astropulse, is taking place at the world's largest radio telescope at Arecibo in Puerto Rico, a long-time SETI workhorse, made famous by the movies *Contact* and *GoldenEye* (see Plate 13). Although this instrument has much greater sensitivity, it has a very small field of view. These projects are a beginning, but a thorough search for alien beacons remains stuck at the planning stage.

NARROWING THE SEARCH

I began this chapter with a plea to widen the search for extraterrestrial intelligence. But a completely unfocused approach is unlikely to succeed, given the needle-in-a-haystack nature of the enterprise. In the case of beacons, the task is made less onerous by concentrating on the regions of the galaxy where most stars are located. The structure of the Milky Way resembles a flat disk with spiral arms protruding: one of those arms contains our solar system. The outer regions of the galaxy are sparsely populated and poor in heavy elements such as life-giving carbon. It is the inner regions that have most of the stars, especially older ones – the ones most likely to have ancient civilizations near them – so the best hope of spotting a beacon is to look in the direction of Sagittarius, where the galactic centre is located.[10]

The radial direction is only half the story. What about habitability as a function of distance 'up' and 'down' from the galactic plane? This is a more complicated topic, because stars migrate up and down in the transverse direction as they orbit the galaxy. The sun, for example, performs such an oscillation once every 62 million years, wandering some 230 light years out of the plane as a result. A few years ago two Berkeley physicists, Richard Muller and Robert Rohde, made an astonishing discovery when looking at fossil evidence for marine life over the past 542 million years.[11] It is well known that the abundance of life on Earth undergoes sharp variations owing to sudden mass extinctions. There are many theories as to why these grisly exterminations occur: for example, cosmic impacts, supernovae, runaway volcanism. What Muller and Rohde found was a distinct 62-million-year cycle in the pattern of marine extinctions, with the death rate highest when the solar system is located at a maximum distance from the galactic plane in the direction of (galactic) north and lowest when it is down south. Their analysis suggests the presence of something nasty beyond the northern edge of the galaxy. What might it be, and why isn't it found on both the north and south sides? (If it was, there would be a cycle of 31, not 62, million years.)

An intriguing explanation has been provided by two University of Kansas astrophysicists, Mikhail Medvedev and Adrian Melott.[12] They point out that although the bright disc of the Milky Way is symmetric between north and south, the galactic halo isn't. The galaxy emits a wind in the form of protons and other charged particles, creating a tenuous cloud that extends far out into intergalactic space in all directions, but configured to be lopsided towards the south. There is a good reason for this. The Milky Way, along with other galaxies in our neighbourhood, is hurtling at 200 kilometres (125 miles) per second in the direction of a massive cluster of galaxies in the direction of Virgo – which lies due north, galactically speaking. The even more tenuous intergalactic medium (consisting mostly of ionized hydrogen gas) serves as a viscous impediment, and this has deformed the halo towards the south, creating an asymmetry. Where the halo gas meets the intergalactic medium, a bow shock is created. Over time, the energy in this shock front gets transferred, via a magnetic process, to protons from both the intergalactic medium and the halo, accelerating them to very high energies. It is these protons (plus others accelerated in a similar manner on the edge of the halo) that make up a large fraction of the higher-energy cosmic rays hitting the Earth. Our planet is protected somewhat by its own magnetic field, but also by the magnetic field of the galaxy. What Medvedev and Melott concluded is that the intensity of this cosmic radiation as received by Earth is surprisingly sensitive to the solar system's location. When it is 'up north', closer to the shock front, the high-energy cosmic ray flux is some five times greater than when it is 'down south'.

Cosmic rays have long been implicated in species extinctions. A high cosmic ray flux hitting the upper atmosphere creates chemical changes that can increase cloud cover – perhaps triggering dramatic global cooling. It can also create a rain of damaging subatomic particles called muons that penetrate deep into the oceans to menace marine life. On top of this, cosmic rays attack the ozone layer, letting in deadly ultraviolet radiation from the sun. The combined effect is to compress the zone for intelligent life to a band away from the north side of the galactic plane. It is unlikely that a technological civilization would evolve on an Earth-like planet too far on the north side, although an advanced civilization that formed before the host star system migrated north may have the know-how to 'batten down the hatches' for some millions of years and ride out the cosmic ray storm.[13] Most long-lived civilizations, however, would be expected to arise around stars that perform smaller-amplitude oscillations and remain close to the safe region of the galactic plane. It would make sense for an alien civilization using beacons to slash costs by concentrating the beam in this 'life plane' of the galaxy, rather than blasting the ether in all directions indiscriminately. Consequently, if beacons are out there, they should appear to us to be clustered in this plane.

Alien civilizations could make use of natural beacons as markers, in the expectation that radio astronomers on other planets would be studying these objects anyway, and might notice if there was something odd about them. Zooming in on these objects specifically would help us narrow the search still more. Pulsars are powerful radio sources familiar to astronomers, and could be used to attract attention to an artificial signal. A pulsar is a spinning neutron star[14] that sprays out charged particles, which then emit an intense narrow beam of radio waves. As the star rotates, so the beam sweeps around – just like a lighthouse. From Earth, the phenomenon is perceived as a highly regular series of radio pulses. Some neutron stars spin so fast that the pulses are spaced by only a few milliseconds. These objects are of great interest to astronomers and much studied. William Edmonson and Ian Stevens of the University of Birmingham in the UK have suggested that aliens might try transmitting artificial bleeps in the direction of habitable planets that lie close to their line of sight of a pulsar, and do so with the same pulse rate.[15] If Earth was one of the target planets, we would pick up these distinctive pulses from a direction in the sky *opposite* to that of the pulsar, which is a dead giveaway for something intelligent and artificial. Edmonson and Stevens have identified a few dozen potentially life-supporting stars that lie within cones of 1° on the side of Earth facing away from highly stable, rapidly spinning pulsars. They have also compiled a list of likely stars in the forward direction, i.e. closely aligned with the pulsars. Because the signal would consist of regular beats with a known period (that of the pulsar), a much weaker signal could be spotted amid the background radio noise, by integrating the observations over a long duration. A more technologically savvy civilization might try using the pulsar emission itself to convey the message, by modulating the natural pulses in some way. That would neatly solve the power problem – pulsars are so powerful they can be detected across the entire galaxy with a modest radio telescope. The signal would then show up as a pattern in the frequency, intensity or polarization of the radio pulses.

A beacon that just goes bleep would of course be of limited value to the transmitting community, because a transient pulse is by its very nature unable to encode a large amount of information. It could, however, serve as a key, enabling access to a much larger database. The beacon could, for example, indicate how to download *Encyclopedia Galactica* from a repository. But where might the nearest repository be? Half-way across the galaxy? Maybe. But there are also reasons why it might be right on our own astronomical doorstep.

A MESSAGE ON OUR DOORSTEP

The biggest drawback of conventional SETI is the immense time required for radio signals to pass between the stars. If we did discover another civilization 1,000 light years away, it would take at least 2,000 years for us to receive a reply to any message we might send them. As Carl Sagan once remarked, that hardly makes for a snappy conversation. Viewed on a geological or evolutionary timescale, two millennia may be the blink of an eye, but in human terms it is dispiritingly slow. But there is another, more exciting, possibility. Humans could conduct a conversation with an alien intelligence by proxy on a nearly real-time basis if the aliens have sent a probe to the solar system, where the travel time for signals to Earth is measured in minutes or hours.[16] Ronald Bracewell raised this possibility at the inception of SETI, and it has been a recurring theme ever since.[17]

From the standpoint of the aliens, the big plus of a probe is its 'set-and-forget' character. With careful design, it might well outlive the civilization that launched it. It doesn't need a massive antenna, unless required to report back to HQ on the home planet. Radio telescopes on Earth had no trouble picking up the Pioneer 10 spacecraft at the edge of the solar system (before it finally blinked off the air a few years ago), and its transmitter was no more powerful than a Christmas tree light bulb. An alien probe could store a huge amount of information in a tiny chip; once in communication with us, its supercomputer could engage in an intensive educational and cultural exchange. In principle, the probe could be any size at all, but for now I have in mind something the size of a human communications satellite.

Would we know if there was an alien probe in our vicinity? Where should we look? The easiest set-up from our point of view would be a probe in low orbit around Earth. However, this can be ruled out: the plethora of orbiting material – most of it human space junk – has been pretty thoroughly catalogued, and there are no unaccounted for objects circling above our heads. What about farther out? A small probe in geosynchronous orbit[18] (which is much higher), or circling the Moon, would probably have escaped our attention so far. Newtonian mechanics shows that long-term stable orbits are rare and must be chosen with care to avoid the need for frequent orbital corrections. Fortunately, there are two points in space where the gravitational fields of the sun and Earth conspire to create stable orbits that keep step with Earth as it goes round the sun; these are known technically as L4 and L5 Lagrange points. SETI scientists are on to this; several preliminary searches of the Lagrange points have been made, but have not thrown up anything unusual.[19] What hasn't been tried, as far as I know, is beaming strong radio signals from Earth to L4 and L5 as a means of 'waking up' a dormant alien probe that might be parked there.

The rest of the solar system is so vast that a systematic search for a small probe is completely unrealistic. An artificial object in the asteroid belt, where it would be surrounded by rocky debris of all shapes and sizes, would be almost impossible to spot, especially if it was anchored to an asteroid. A precisely spherical or conical shape, or a collection of objects connected by struts, would obviously make us sit up, but if the aliens wanted to deliberately conceal a probe, it would be easy enough to do. Clearly, there could be a large number of alien probes in the solar system, and we would be completely unaware of them unless they signalled us.

There is no reason why a probe should have arrived in the solar system only recently. It could have been dispatched millions of years ago by a civilization that had determined, using remote observation, that there was life on Earth. The probe would remain passive, quietly monitoring our planet and biding its time until a technological society emerged. At that point – if the probe's computer thought it prudent – it could initiate contact. How would that happen? The obvious method would be for the probe to send us a radio signal. For us to recognize its exceptional nature, the signal would have to grab our attention as something very much out of the ordinary. One suggestion (used by Carl Sagan in *Contact*) is that the probe beams back to us an early radio or TV broadcast. It would certainly strike us as baffling if a radio telescope detected a broadcast of *I Love Lucy* coming from deep space. (For the record, the first episode of *I Love Lucy* was broadcast on 15 October 1951.) On the other hand, if such a show were picked up by domestic TV sets, viewers wouldn't think it at all odd – the show would just be dismissed as another network repeat.[20]

A more far-out proposal is that the probe might make use of the internet to communicate with us. The probe's on-board computer would doubtless be programmed to first assess the level of development and the general character of human society before deciding to disclose its presence. What better way to build up a picture of humanity than by monitoring websites, e-mail messages, chat rooms, YouTube, etc. After all, that is exactly what government spying agencies already do. When the time is ripe, the probe would then log on to an appropriate website via a microwave link and publicly announce its existence.

A group of SETI enthusiasts led by a Canadian researcher, Allen Tough, have taken the idea seriously enough to set up a dedicated website inviting ET to log on (http://www.ieti.org/). The reader who takes the trouble to look will find my name as one of the signatories supporting this admittedly eccentric but delightfully imaginative project. Understandably the website attracts a steady stream of clever hoaxers, but, alas, no extraterrestrial probes – so far at least. The existence of the website does, however, raise the thought-provoking question of just how one could be sure that a contactee really is an alien entity rather than a human prankster. It would be terrible if ET called and we responded by saying 'pull the other one'. A few years ago, Allen telephoned me about an intriguing contender who had swiftly passed a number of basic tests designed to filter out crude hoaxes. He asked me to suggest a sure-fire way of spotting a fake. I suggested that he send back a hundred-digit number composed of the product of two primes, and ask the contender to factor it back to the original. The point here is that multiplying numbers is easy, but going the other way – factoring – is much harder. By way of illustration, most people would take less than a minute to work out, say, 141×79 = 11,139, but if you are asked to find two prime numbers which, when multiplied, yield 11,139, it will take far longer. In effect, you have to run through all the possibilities and eliminate them one by one until you hit on the right answer. A computer faces the same obstacle, and for seriously large numbers even the fastest supercomputer in the world is flummoxed. For that reason, the product of prime numbers forms the basis of most encryption techniques. Allen duly came up with some numbers and, to our surprise, the contactee delivered the correct answer in pretty short order! So we tried a 200 digit number, which we knew to be (at that time) beyond the performance capability of any known human supercomputer. At this point the hoaxer, a bored computer operator in Birmingham, UK, threw in the towel. The problem with the prime-number test is that it could be defeated by a quantum computer, should one ever be built (see Chapter 8). So far, in spite of millions of research dollars, quantum computation remains in its infancy. But if a functioning quantum computer is made one day by humans, we shall have lost a very useful discriminator of extraterrestrial technology.

Another popular idea is that an alien artifact may have been placed on Earth itself. If so, would we have found it? There are plenty of places such an object could lie undetected – the bottom of the ocean, say, or buried deep in the Greenland icecap. It could lie just below the ground on almost any place on Earth without having been spotted. All these scenarios have been used in science fiction, but it's not clear why an extraterrestrial civilization would deliberately conceal an artifact in this way.

If aliens sent a probe here entirely on spec, without knowing whether Earth has, or might one day have, a technological civilization, then there is a high probability that it arrived a long time ago – say 10 million years, or even more. A major problem facing the probe's dispatchers would be to create an artifact that could remain intact and functioning for such an enormous length of time. (Our own technology remains functional for only a few decades.) From the point of view of durability, the surface of the Earth is an unpromising location to park a probe, because of geological upheavals such as glaciation, comet impacts, volcanic eruptions, earthquakes, etc. A less variable location is the Moon, if the object were buried deep enough to avoid small meteor impacts. That scenario was explored by Arthur C. Clarke and Stanley Kubrick in the famous story *2001: A Space Odyssey*, where the alien artifact is depicted as a giant obelisk. Although the Moon's surface has been photographed pretty thoroughly, if the probe were small, or buried, we wouldn't yet know about it.

NANOPROBES, VIRAL MESSENGERS AND GERRYMANDERED GENOMES

One objection to 'spreading the word' with high-speed probes as opposed to radio signals is cost. For example, a one-ton spacecraft travelling at a modest one-tenth of the speed of light, would require half a billion billion joules of energy to launch, equivalent to the Earth's entire power output for several hours. And this ignores the need for the probe to – somehow! – slow down on arrival, which might require the same or even more energy. There would have to be very strong motivation to embark on such a project out of altruism or curiosity (as opposed to desperation, e.g. to preserve something before Armageddon struck), especially if it entailed dispatching an entire fleet of probes to cover a wide swathe of the galaxy.

Fortunately there is a way to cut the energy factor dramatically, by building smart probes that can self-repair and reproduce themselves as they go. Then instead of ET sending a probe individually to every promising star system, a single probe could be dispatched and left to multiply. The concept of a self-reproducing machine was first explored by the Hungarian mathematical physicist John von Neumann, who, along with the English mathematician and Second World War code-breaker Alan Turing, is credited with the invention of the modern electronic computer (thus finally realizing Babbage's nineteenth-century conception). A computer is a universal machine, in the sense that a single device can be programmed to solve all computable problems. The concept of a universal computer leads very naturally to that of a universal constructor – a machine able to make other machines according to an internal program. Suitably programmed, a von Neumann machine could also make copies of *itself* (including the copying instructions), and would therefore constitute a self-reproducing machine.[21]

It is easy to imagine an advanced civilization sending out von Neumann probes to explore the galaxy. On arrival in a star system, one such machine would mine raw materials from asteroids or comets in order to replicate. Some of the progeny might then study the planets, and perhaps try to contact any intelligent life, beaming back information to the home planet. They might even stay on indefinitely in the star system to serve as beacons, or as silent probes, while others travel to the next star system. The process could continue ad infinitum, with the total number of machines rising exponentially. In this way, the building costs of the exploration programme would not all fall on the originating civilization.

There is scope for further dramatic improvement in cost by

miniaturization, dispensing with fancy equipment and radio transmitters. If the purpose of the probes is merely to disseminate a message, or basic information about the dispatcher, then there is a much easier way to go about it, which is to use nanotechnology. In 1959, the same year as Cocconi and Morrison published their visionary paper about SETI, a no less visionary lecture was delivered by Richard Feynman, the brilliant and creative theoretical physicist. Entitled 'There is plenty of room at the bottom', the lecture foreshadowed molecular-scale engineering decades before it came to fruition. Today, nanotechnology is advancing rapidly. First there was the incredible shrinking microchip, then the scanning tunnelling microscope capable of moving individual atoms in a controlled way, then carbon nanotubes and quantum dots. Nanotechnology is likely to have a spectacular impact on information storage. In a January 2000 address on science and technology, President Clinton discussed the US National Nanotechnology Initiative and referred to some of the possibilities, such as 'shrinking all of the information housed in the Library of Congress into a device the size of a sugar cube'.[22] It has been estimated that the contents of a substantial encyclopedia could be packed into a volume smaller than a bacterium. Progress is so rapid that alarmists are predicting the end of the world as we know it, with runaway nanomachines transforming the surface of the planet into 'gray goo'.[23] Strictly speaking, 'nano' refers to a scale of size one billionth of a metre, corresponding to a large molecule, but the term is used more loosely to refer to all ultra-small-scale engineering.

In the not too distant future, when humans will be able to build micro- or nanomachines that store prodigious amounts of information, they could be used as space probes. Because of their tiny size they could be accelerated to high speeds (say 0.01 per cent of the speed of light) very cheaply, perhaps without the need for rockets. It may still take a few million years for them to reach the target stars, but haste is not an issue in the scenario I am exploring. We can readily imagine an advanced alien civilization packaging mini-databanks in microscopic capsules and spewing them around the galaxy in the millions.

A nanoprobe differs from the Bracewell-type probe I discussed earlier, in that it couldn't send out radio signals to attract attention. How, then, would it make an impact? This is where von Neumann's idea comes in. If the nanoprobe were a self-reproducing von Neumann machine, then on arrival it could replicate like crazy until its progeny formed a conspicuous scum that a curious scientist might analyse under a powerful microscope. There is a more elegant strategy however. Nature has already invented neatly packaged data-rich nanomachines: we call them viruses.[24] A typical virus contains thousands of bits of information encoded in either RNA or DNA – enough for a decent message. So why not engineer trillions of viruses, package them in pea-sized microprobes, and spew them around the galaxy? Each virus would convey a message for any future intelligent life on the destination planet,[25] the space age equivalent of a message in a bottle. The beauty of the scheme is that the message can be replicated ad infinitum should it encounter life on a destination planet, by the simple expedient of programming the viruses to 'infect' any DNA-based cells with which they come into contact. The virus inserts its message into the genetic material of the host organism's germ cells (that's what so-called endogenous retroviruses do), and the cell obligingly replicates it and passes the message on to all future generations. In this way the virus would spread like wildfire through the host ecosystem, its information preserved for millions of years until some future Craig Venter begins sequencing genomes and stumbles across the message. Certainly DNA does get inserted into living cells in this manner; whole chunks of human DNA are the genomic detritus of ancient viruses that infected our ancestors.

The way I've described it makes it sound simple, but in reality some major technical hurdles stand in the way. Most obvious is that DNA may be only one of many ways that biological information is encoded, and it is hard to see how the aliens would know in advance what terrestrial life uses. A second problem has to do with physics. Interstellar space is a dangerous environment. Cosmic rays in particular can cause serious damage to nanostructures, and in time they would break up the molecular message. Shielding would ameliorate this problem, but at the expense of adding mass. In addition, the projectile has to be slowed on arrival to enter the atmosphere of the target planet without incinerating itself. Carrying fuel to decelerate would also add – very substantially – to the payload mass. These refinements would scupper the small, fast and cheap philosophy behind the idea of microprobes. Possibly the technical problems could be solved without adding lots of extra mass – for example, by using aero-braking for deceleration – but even if they could, the engineered viruses would face serious biological issues on arrival. Viruses are highly attuned to their hosts, which is why you can swim in the sea – virus soup, remember – and not get sick (mostly). So even if ET guessed that Earth was replete with DNA-based life, without knowing the specifics of the host genomes it's not clear how a virus could be designed to work reliably. Perhaps universal, or general-purpose, viruses can be made, which infect a range of organisms without killing them.

A second problem concerns mutations. Once the message has been inserted, it needs to remain unchanged for as long as possible to stand a good chance of being discovered one day. But natural mutations occur all the time during the DNA copying process; and a mutated message is a scrambled message – sense degenerating into nonsense. Natural selection can serve to stabilize genetic information, but only if there is selection pressure, in other words, if the mutation has detrimental consequences for the survival of the organism, and gets weeded out of the gene pool. If the inserted segment – the message – is biologically inactive (i.e. if it's just being carried along for the DNA ride) – it's hard to see how natural selection would operate to conserve it. A lot of DNA seems to be 'junk' – great sections that don't code for anything, and so mutates rapidly and harmlessly over the generations, unchecked by selection. Assuming the viral DNA is treated by the host organism as just more junk, the message risks being garbled by mutations after a few thousand generations. Recently, however, some doubt has been cast on this simple picture. Substantial sections of what appear to be identical sequences of junk DNA have been found in both human and mouse genomes, suggesting that these sequences have been conserved since pre-mice and pre-humans parted genetic company 40 million years ago. Now maybe these sequences fulfil some vital role in a subtle way, but it's not obvious: when they are deleted from the mouse genome, the mice seem perfectly happy. So it's possible that sections of junk can be accurately replicated and conserved for millions of years, perhaps by somehow chemically piggy-backing on key genes that are under strong selection pressure, and so conserved. Anyway, if an alien virus insinuated itself into the host genome in such a piggy-back manner, the message could be good for tens of millions of years.[26]

There is an alternative way to deliver a biological message that avoids some of the problems with viruses. Rather than trying to hijack indigenous life, the aliens could try to create an artificial shadow biosphere *ab initio*. A civilization a few thousand light years away could, even from that distance, know enough about Earth's geology, atmosphere and chemical composition to deduce something about our biology and environmental conditions. Armed with that information, they could design novel microbes customized to flourish in the terrestrial environment, living peacefully alongside indigenous organisms. The synthetic cells need use neither DNA nor proteins, and could be designed to thrive in conditions too extreme for Earth's indigenous life, thus avoiding direct competition. By using molecular structures with stronger bonds than DNA the cells would suffer less cosmic ray damage en route. The all-important message sequences would be carefully engineered so as to mutate only very slowly, possess in-built redundancy and enjoy error-correcting mechanisms of the sort employed by terrestrial organisms. The package of microbes would be targeted at Earth specifically, or any other planet likely to spawn intelligent life one day. On arrival, the microbes would take up residence, spread across the planet, possibly adapting to changing conditions, and hang out innocuously for tens of millions of years awaiting discovery. If we ever do detect a shadow biosphere, it would be a more plausible place to look for an alien message than in the genomes of life as we know it.

The feasibility of using microbial cells to send messages between the stars hinges on whether they can be delivered efficiently. Michael Mautner, a New Zealand chemist who also runs something called the Panspermia Society, has done some calculations to find out. He believes it would work. In fact, he thinks humans could do it with foreseeable technology. The key is to microminiaturize the payload. Mautner envisages centimetre-sized membranes with tiny pellets embedded. The microbes ride inside the pellets, along with a starter kit of nutrients. The membranes reflect the solar wind and the light of the sun, thereby receiving a small but persistent propulsion force. Accumulated over years, this tiny effect could gently accelerate the capsule to 0.01 per cent of the speed of light. Once the diminutive spacecraft reaches

1. Part of the SETI Institute's Allen Array in northern California, showing two of the many linked antennas.

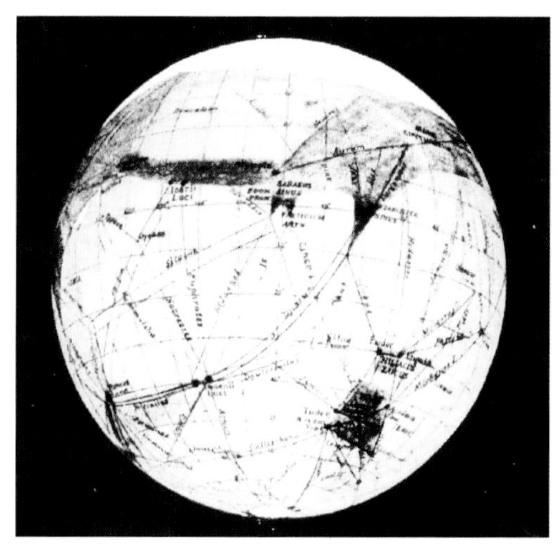

2. The canals of Mars, according to Percival Lowell.

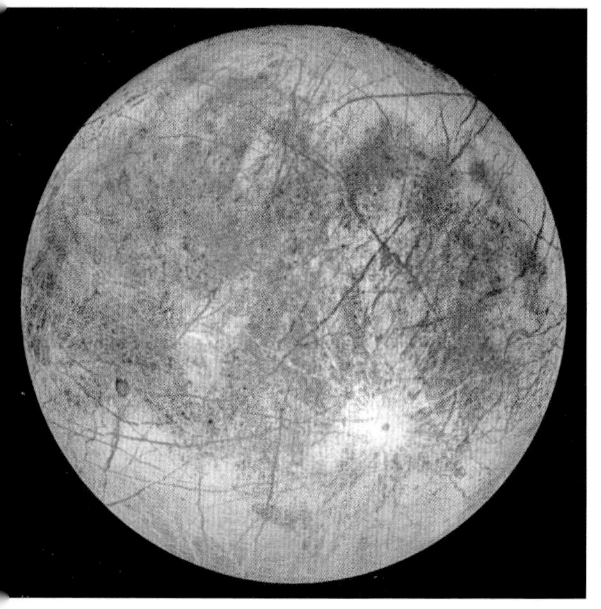

3. Europa, a moon of Jupiter, showing an ice-covered surface rent by striations thought to be caused by slippage of the ice on a subsurface liquid-water ocean.

4. Viking spacecraft, showing the robot arm used to gather dirt for biological analysis.

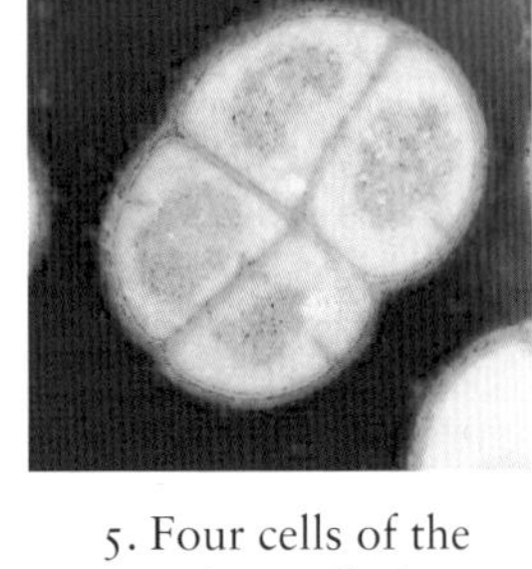

5. Four cells of the amazing radiation-tolerant *Deinococcus radiodurans*.

6. A submarine volcano located on the Juan de Fuca Ridge in the North-East Pacific. The 'black smoke' is a turbulent cloud of iron sulphide particles.

7. The dry core of the Atacama Desert, where even the hardiest known microbes grind to a halt. This region might just be home to weird life.

8. A piece of the Murchison meteorite, which contains amino acids, the building blocks of proteins.

9. This Mars meteorite, found in Antarctica in 1984, contains tiny features (see inset) suggestive of nanobacteria.

10. Aliens in the lake? Felisa Wolfe-Simon and Ron Oremland search for arsenic life in Mono Lake, California.

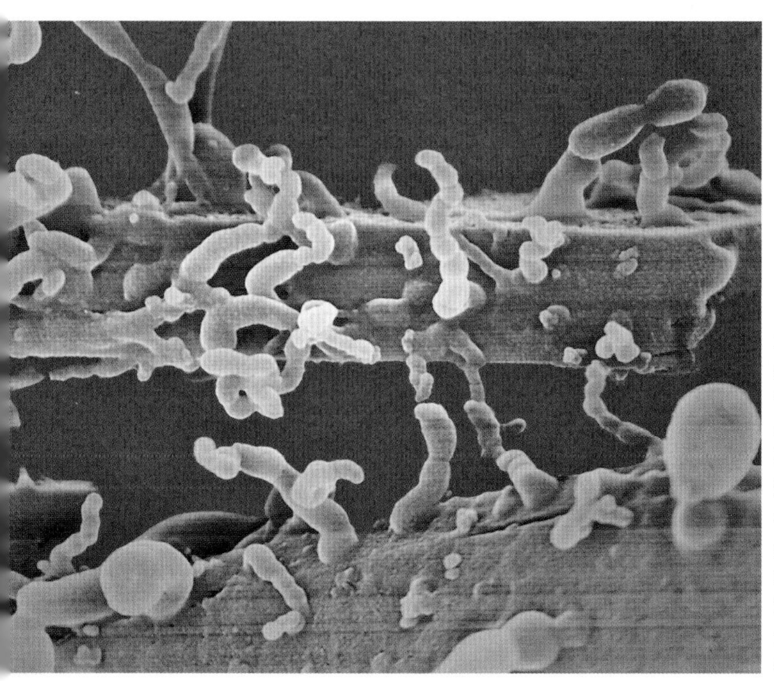

11. These minuscule shapes, dubbed 'nanobes' by their discoverer, Philippa Uwins, have been interpreted by some – controversially – as a weird form of life. They are too small (about 100 nm) to be standard microbes.

12. The radio telescope at Parkes in New South Wales, Australia, has been at the forefront of SETI research. It is one of the most powerful radio telescopes in the world, and was used to relay the first moon walk in 1969, an event made famous by the movie *The Dish*.

13. The Arecibo radio telescope is the world's largest but is not steerable, so it can observe only a limited slice of the sky. It has for several years been used intermittently for SETI.

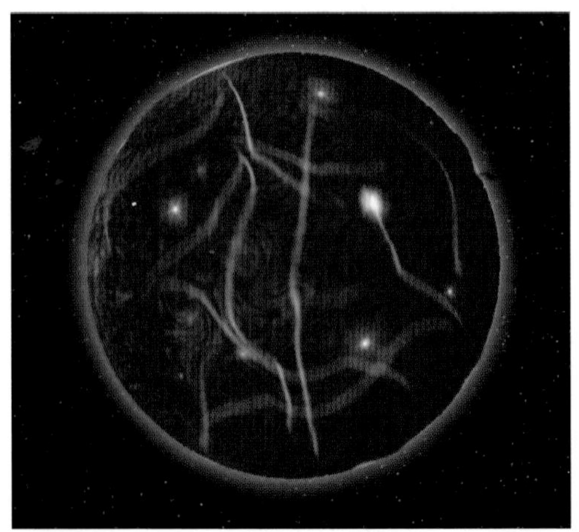

14. Matrioshka brain. Is this the real ET?

cruising speed, the solar sail could detach, or fold up around the pellet for added protection against cosmic rays. For most of the journey not much would happen. The microbes would simply lie dormant, the pellet would cool to a few degrees above absolute zero and the little bag of tricks would whiz unobtrusively across the interstellar void. On approach to the target planetary system, the pellet would fragment, turning a speeding bullet into spreading buckshot. Mautner has calculated that a speck 60 micrometres across could survive aero-braking into a planet's atmosphere without incinerating its cargo.

A different strategy would be for the aliens to use comets as delivery vehicles. Following a series of clever gravitational deflections, a comet could be flung out of the aliens' planetary system towards ours. There is good evidence that dormant microbes or viruses could survive inside a comet for many millions of years, which is certainly long enough to traverse light years of space at typical ejection velocities. When a comet comes close enough to the sun, it begins to evaporate, sprouting a characteristic tail as gas, water and microscopic particles stream off. If the comet were laden with engineered bacteria, viruses or some other type of microbiological entity, they would spew forth too, forming a long, diffuse infectious cloud. Should it happen that the Earth sweeps through such a cloud, it would acquire a dose of viable biological agents.[27]

However speculative the idea of 'genomic SETI' might be, it makes sense to take a look for gerrymandered genomes. And that just what Hiromitsu Yokoo and Tairo Oshima of the Kyorin University Hachioji Medical School in Japan did as long ago as 1979. They searched the DNA of ϕX174, a bacteria-infecting virus known as a phage, to see if it contained anything fishy.[28] It didn't, but that was in the early days of bioinformatics. Today, genome sequencing is a major industry, with many organisms, from microbes to humans, having their DNA read and posted on the internet. The time is ripe to do a systematic search of these genomes to look for arresting oddities. The sequencing is being done anyway, so it costs almost nothing to run the data through a computer to look for suspicious patterns. In fact, the highly successful SETI@home project was emulated by genome@home, now sadly suspended. It would be simple enough to merge the two. Who knows what might come out of it? The project could paraphrase the *X Files* and be promoted with the catchy slogan: 'The Truth is in There'.

6

Evidence for a Galactic Diaspora

When you have eliminated the impossible, whatever remains, however improbable, must be the truth.

Sherlock Holmes[1]

WHERE IS EVERYBODY?

In the summer of 1950 the Italian physicist Enrico Fermi was working at Los Alamos in New Mexico, at the research laboratory where the atomic bomb was designed during the Second World War. Fermi was by then a legendary figure in theoretical physics, having solved many problems in quantum mechanics, particle physics and astrophysics, as well as playing a central role in the Manhattan Project. He was regarded, in fact, as the archetypal genius (see Fig. 9). One day Fermi was strolling to lunch with some colleagues, including Edward Teller, often called the father of the H-bomb, and John von Neumann (whom I mentioned in the previous chapter in connection with self-reproducing machines) when the conversation turned to UFOs, or 'flying saucers' as the press had dubbed them, which were being reported in large numbers at that time. This naturally led to a lively discussion about the probability of extraterrestrial life and the likelihood that flying saucers were in fact alien spacecraft. In the midst of the debate, Fermi suddenly asked, 'Where is everybody?', referring, of course, to the putative alien beings. If the galaxy really is teeming with life, he explained, then Earth should have been colonized in the far past. The aliens ought to have been here all along, and we would be well aware of it.

Fermi's basic argument is simple enough. Life on Earth has taken

Fig. 9. Italian genius Enrico Fermi.

3 or 4 billion years to evolve to the level of intelligence and technology. If life started on another planet, say X, at the same time as it did on Earth, the probability that life on X would attain the same level of technology as humans at this particular time, even to within a few thousand years either way, is exceedingly small. Consider the many chance events that have occurred over billions of years of evolution, such as the dinosaur-destroying impact 65 million years ago. What are the odds that a similar impact would have occurred and wrought a similar transformation on planet X, at roughly the same time? Negligible. If X evolved intelligent life and technology by some other evolutionary pathway, then it might reach the level of human technology tens or even hundreds of millions of years earlier. Or later. If Earth were typical, and if there are lots of Planet Xs out there, then life on some of them will evolve intelligence more slowly than here; those planets will not attain technology for a very long time yet. On others, the evolution of intelligence and technology will proceed more rapidly, so that they will have reached our level long ago, perhaps 100 million years or more. Now add the fact that there were Earth-like planets before our solar system even existed: on those planets, life would have a huge head start over Earth. Putting all this together, the conclusion is clear: if life is widespread and Earth is typical, there should have been many planets with advanced spacefaring civilizations long, long ago. So why haven't the aliens come here already? This, in a nutshell,

is what has become known as 'Fermi's paradox'. Strictly speaking it is not a paradox in the philosopher's meaning of the term, but simply an unavoidable consequence of some fairly plausible assumptions. So what is the answer?

The most obvious explanation for the absence of aliens on Earth is that aliens don't exist – that is, we are alone in the universe. That was presumably Fermi's own position, and the point of his argument was to pooh-pooh the flying-saucer stories. If that is the correct answer, then SETI is a waste of time and money. But we mustn't be too hasty in drawing this pessimistic conclusion. There could be any number of reasons why alien civilizations are out there, but not here. An entertaining book by Stephen Webb lists no fewer than fifty explanations for ET's conspicuous absence,[2] ranging from the 'zoo hypothesis' (we are being watched, but not contacted) to the 'parallel universe' hypothesis (the aliens are having too much fun exploring other universes to bother with us). Take your pick.

By way of illustration, consider the following resolution. Suppose there are many civilizations in the Milky Way, and they long ago established a galactic network of information exchange. This is an idea dating back to 1974, when the Stanford University astronomer Ronald Bracewell envisaged a 'Galactic Club' of communicating civilizations, sharing news, information and gossip, with data zipping from star to star like e-mails over a cosmic internet.[3] The club might even have been established before the solar system formed, 4.5 billion years ago (the galaxy is over 12 billion years old). Some members would drop out as their civilizations faded or were destroyed by a catastrophe, others would sign up as they attained radio technology and discovered that there was a network of information exchange already operational. Bracewell regarded humanity as on the verge of joining this Galactic Club – as its newest member – a step that would bring us untold benefits, but would also serve as a strong disincentive to embark on interstellar travel. If the motivation to explore is curiosity and information-gathering, it is far easier to simply log on to the GWW (Galactic Wide Web) and obtain the information for free. It is, after all, much faster and cheaper to send radio waves across interstellar space than big metal machines. If there is somebody at the destination planet already, then why bother to make the trip? If the purpose of space travel is exploration, well, the aliens can send us the content of their latest DVD. On the other hand, if it is conquest, then the fact that the target planet already has a far more advanced civilization ensconced would constitute a pretty strong deterrent. All in all, it would make more sense for the newcomer civilization to stay put and simply join the Galactic Club. But if nobody is travelling, there is no reason why the aliens should be here, or should ever have passed this way. It doesn't mean there isn't anybody out there, only that space travel is not an idea with enduring appeal. I believe this argument has some force, but it is convincing only if there is a very large number of planets with indigenous technological communities. If there is plenty of untouched planetary real estate to go round, then a civilization might well move to occupy it, even while remaining in 'the Club'. Also, it is important to guard, as always, against anthropocentrism. Humans have been keen to migrate for reasons of curiosity, material gain or conquest. But there might be many motives for an alien civilization to expand into space, some of which would mean little to us.

One issue that isn't relevant is the enormous distances between the stars. It's true that it would take a long time by human standards to complete the journey from one star system to another, even for a very high-speed craft. However, at a tenth of the speed of light, only a million years is needed for a spacecraft to cross the galaxy. If there were an alien civilization anywhere in the galaxy during the past, say, one billion years, the million-year journey is well within its time frame. Of course, it may not want to make the trip in one great leap. Most likely it would go from one planet to a nearby one, perhaps in large space arks that take many generations to complete the trip, and take up residence on each. Eventually a settlement would mature, and the colonists would venture on to the next suitable planet, and so on. This creeping colonization is slower than an expedition targeting a specific destination planet, but not by much on an astronomical timescale. If it took 1,000 years for the colony to mature, and suitable planets were situated, say, an average of ten light years apart, then the accumulated planetary sojourn time would add only about 3 million years to the total time needed to reach Earth from the inner region of the galaxy, where the older stars reside and where the most advanced civilizations would presumably be located. So that's less than 4 million years to get here, all told. Of course, one wouldn't expect the aliens to make a beeline for Earth, given the rich pickings of all those other habitable planets on the way. Rather, we can imagine the seed civilization spreading out its colonizing tentacles in all promising directions, perhaps to engulf the entire galaxy eventually. A diffusion process like that would take longer, but it would still constitute only a small fraction of the age of the galaxy. Obviously not every spacefaring civilization would choose to colonize the galaxy in a grand imperial manner; and it had better not, or there would be unpleasant clashes all the time. But it takes only *one* such community somewhere in the galaxy to present us with Fermi's awkward conundrum.

When Fermi stated his original 'paradox', he had in mind flesh-and-blood aliens coming to Earth, but the same reasoning also applies to alien artifacts, especially if they are capable of multiplying and spreading, like von Neumann machines. When it comes to space exploration and colonization, self-reproducing machines offer many advantages over biological pioneers in cost, durability and survivability. If extraterrestrial civilizations are common, surely the galaxy should already be overrun with von Neumann machines, because they could colonize the entire Milky Way in a time much less than the age of the solar system. As no evidence has (yet) been found for von Neumann machines in our astronomical neighbourhood, their absence could be taken as tipping the scales against the hypothesis that extraterrestrial civilizations are commonplace.

The physicist Frank Tipler has argued forcefully that the apparent absence of von Neumann machines in the solar system all but proves we are alone in the universe. He estimated it would take only 300 million years for the galaxy to be flooded with these devices, so there has been plenty of time for a galactic takeover to happen. Tipler reasons that von Neumann probes are a highly effective form of interstellar migration, on both logistical and economic grounds, and therefore their absence represents a more potent version of the Fermi paradox. It is easy to think up reasons why living beings might avoid travelling between the stars (it's a long way after all); it's less easy to understand why alien von Neumann probes wouldn't do it.

Tipler's argument works only if we accept his major premise, which is that there *are* no von Neumann machines in the solar system. Can we be sure of that? Obviously we can rule out the scenario in which alien von Neumann machines just go on multiplying until they overrun the solar system. But for a less aggressive strategy, the situation is not so clear-cut. As I explained in the previous chapter, there are countless places that a small inert machine could be skulking, unbeknownst to us. Still, it's hard to understand the purpose of such a programme, if it is not to establish contact with indigenous intelligent life. In which case, why the eerie silence?

AND WHERE ARE ALL THE TIME TOURISTS?

There is a curious temporal version of the Fermi paradox, articulated most famously by Stephen Hawking in 1992, when he asked 'Where are all the time tourists from the future?'[4] Hawking concluded from their absence that travelling from future to past isn't on. It must be admitted that time travel lies on the borderline between science fact and science fiction – a tantalizing dream for which the best one can say is that it hasn't yet been proved impossible. Our best understanding of the nature of time comes from Einstein's general theory of relativity, which does seem to permit journeys both forward and backward in time. In fact, travel into the future is already a done deal. It goes by the name of the time dilation effect, and is readily demonstrated by accurate clocks. All you need to do to reach the future sooner is to move – as fast as possible. For example, at 99 per cent of the speed of light, if you set off now you could reach Earth year 2100 in less than thirteen years. However, given that our best rockets achieve less than 0.002 per cent of the speed of light, human time travel is so far limited to pitiful amounts (microseconds only).

Getting back from the future is a much tougher challenge. Although not strictly forbidden by the general theory of relativity, journeying backwards in time involves exotic super-technology such as wormholes in space. Wormholes resemble black holes inasmuch as they both use gravitation to warp time, but whereas entering a black hole is a one-way journey to nowhere, a wormhole has an exit as well as an entrance, permitting the traveller to fall through it and come out somewhere else. Now for a reality check: whereas black holes really exist, there is no evidence whatever for wormholes.[5]

To turn a wormhole into a time machine requires imprinting a time difference between the two mouths of the hole, which entails some tricky manipulations. It turns out that the time required to complete the imprinting process is always longer than the duration of the time difference achieved. For example, it would take more than a hundred years to create a time machine that can access a hundred years of the past. Obviously, then, you can't use a wormhole to visit a time before the date of completion of the machine's manufacture. In this respect 'real' time machines differ from H. G. Wells's fictional version. The absence in 2010 of human time tourists from Earth's future is then perhaps no surprise. However, what if there are aliens with super-technology who already possess time machines? Their descendants could visit us now from the future, or they could lend the time machines to future earthlings and permit them to do 'reality history'. Does the absence of time tourists tell us there are no advanced aliens, or that travel back in time is impossible after all, or that it is theoretically possible, but prohibitively expensive or dangerous? All we can conclude with certainty is that the possibility of time travel only makes the Fermi paradox worse, because it opens up Earth to visits (or invasion) not only from our alien contemporaries, but also from their (and our) descendants. And with time travel, the long journey time between the stars is irrelevant: ET could reach Earth *before* setting out! For those readers interested in learning more about time travel, I refer you to my little book *How to Build a Time Machine*. Fascinating though the subject may be, I will not consider it further in this book – speculating about space travel is already difficult enough.

A COSMIC FOOTPRINT

When contemplating the prospects for human space travel, futurologists split into two camps. One of them predicts a rosy future in which new propulsion systems and economies of scale reinvigorate our push into space. Colonies will be set up on the Moon, then on Mars, perhaps on some asteroids, and new industries will spring up with them, driven by commercial interests.[6] Over the coming centuries, humans will spread across the solar system and beyond, duly fulfilling their cosmic destiny.

The pessimists will have none of this. They see space exploration as an idiosyncratic and transitory diversion rooted in the politics of the Cold War and the urge to seize the 'high frontier'. With launch costs so prohibitive and commercial returns on space flight negligible, the taxpayer will inevitably tire of footing the bill, and the entire space programme will dwindle and peter out. No matter that the scientific pay-off of space exploration is immense; it is an open secret that the US space programme would be scaled back drastically if there weren't substantial military advantages driving it. It's possible to hope, and even expect, a 'new world order' in a century or two that would abolish the military threat from space. If that happens, manned space exploration would be an inevitable victim of the concomitant 'peace dividend'. Signs of waning interest are already evident in stalled budgets for NASA and other space agencies. It isn't hard to convince oneself that no large-scale human presence in space will endure beyond the next decade or so.

I keep dithering over which of the two scenarios – optimistic or pessimistic – I believe. Each is plausible. In terms of the Fermi paradox, however, it boils down to this. Fermi lived at the dawn of the space age, when it was natural to believe that space exploration would be a seamless extension of terrestrial exploration, and would grow exponentially along with science, technology and the global economy. After all, Fermi and his colleagues had just finished building the first atomic bomb. Nuclear-powered rockets seemed a small step away.[7] Flash Gordon, the comic strip hero, ruled the universe. Today, almost five decades after the last Moon landing, space travel doesn't seem quite so inevitable. When reflecting on alien civilizations, it would be rash to conclude on the basis of a few decades of our own space programme that a more advanced civilization will inevitably be spacefaring. However, it would be equally rash to suppose that no alien civilization has ever expanded into the galaxy. Remember that in reflecting on the potential for alien technology, we need to adopt a perspective that encompasses a vastly greater span of time than all of human history.

For fifty years SETI has been motivated by the hope that advanced extraterrestrial civilizations will manifest themselves through their radio emissions. But the eerie silence prompts us to re-evaluate that expectation, and consider other ways an alien intelligence might leave identifiable traces. As every forensic scientist knows, intelligent behaviour can betray itself in many indirect and subtle ways, even when the subjects make a deliberate attempt to conceal their activity. The universe is a rich and complex arena in which signs of alien intelligence might be buried amid a welter of data from natural processes, and unearthed only after some ingenious sifting. Even if we never detect a deliberate signal or beacon from an alien civilization, we might still accumulate enough circumstantial evidence to convince ourselves that we are not alone in the universe.

In order to make progress it is essential to devise strategies that go well beyond traditional SETI. And SETI researchers agree: 'Our experiments are still looking for the type of extraterrestrial that would have appealed to Percival Lowell,' admits Seth Shostak.[8] A comprehensive search for alien technology should involve more than the use of radio telescopes, and preferably encompass the full panoply of modern science, from particle physics, through microbiology to astrophysics. In the broadest sense, alien technology would betray itself through some sort of anomaly, something that 'looks fishy' – out of place or out of context. It might be small, perhaps only a minor perturbation, easily overlooked, but bearing a distinctive hallmark of artificiality. As we don't quite know what it will be, it pays to be as open-minded and imaginative as possible.

Even if we don't know what to look for, we can make some educated guesses about *where* a footprint of alien technology might be found. Fermi ruled out the existence of aliens on the basis of a simple model of migration, in which aliens leave their home planet and spread out uniformly across the galaxy. A more realistic picture of how interstellar migration might play out is to imagine new technological civilizations emerging randomly here and there in the galaxy, some fading away, others enduring, others expanding, the whole process continuing over billions of years. What pattern would emerge? How quickly would the galaxy fill up with migrants? How often would neighbouring civilizations clash or merge? Fermi based his original paradox on an analogy with human migration. Modern humans left their African homeland a little over 100,000 years ago, and quickly spread across the planet, reaching as far as Tasmania, Tierra del Fuego, the Pacific Islands and the Arctic wastelands. The initial step was the colonization of virgin territory. That was followed by a period of consolidation, after which renewed emigration began from the colony in search of more unoccupied land. Step by step this dispersal continued, until all the accessible places on Earth were inhabited. Because the successful roamers lived to spread their genes, Darwinian evolution fixed the wanderlust habit in the gene pool, which is why human beings still feel the urge to climb the next mountain, fly to the Moon or set up colonies on Mars (at least, some of us do), even though for the vast majority of people there is no longer any need to keep moving on in order to survive. Many science fiction writers have extrapolated from history, portraying our descendants reaching for the stars, perhaps establishing a mighty empire, driven to the far reaches of the galaxy by those ancient wanderlust genes and their silent imperative that 'the grass is greener on the other side of the hill.'

But the human experience may be of marginal relevance to alien galactic migration. The motivations of intelligent aliens are a closed book to us. Whatever might induce them to spread out, it is unlikely to be the product of primitive urges that confer little long-term survival value – the relevant genes would, I believe, long ago have been engineered out of the gene pool. When it comes to machine intelligence, we are totally in the dark. Who could guess the strategies that might be programmed into von Neumann probes by an alien mind, or how those strategies would evolve if the self-replicating machines possessed autonomy? All of which makes it hard to figure out under what circumstances an alien civilization would spread into space, and if it did, then in what manner, and how far. Even if the diaspora isn't driven by biological urges ('We gotta get outa this place') it may still be favoured on rational grounds ('A settlement on Planet X would complement our own society nicely'). To model alien migration we have to start somewhere. A good place to begin is with the simple dictum that if something is good, more is better. If a civilization creates something of value on its home planet – a culture, a technological triumph, a grand vision – we don't need to decide what it is – then it seems reasonable that the community would act to replicate it elsewhere. And with that modest investment in assumptions, a surprising amount can be deduced using mathematical modelling.

RIDING THE WAVE

Few would suspect that the humble coffee percolator could inspire an entire branch of mathematics. But percolation theory – so named by analogy with the way that water migrates through coffee grains – has been applied to real-world problems as diverse as hydrology, epidemiology and materials science. It has also been applied to alien migration. The aerospace scientist Geoffrey Landis produced one of the first quantitative percolation models to predict how an alien civilization might spread across the galaxy.[9] Landis made the reasonable assumptions that travel between stars (whether by intelligent organisms, robots or cyborgs) is difficult and expensive, and the number of unoccupied planets suited to colonization is likely to be small. He sensibly rejected the notion of a galactic empire under central control: it takes 100,000 years for a signal to cross the galaxy, so the concept of a unitary galactic culture is ridiculous, however popular it may be with science fiction fans. A more realistic pattern is a patchwork quilt of diverse local cultures emerging as the colonization evolves. Some colonies will be content to consolidate, others will choose to expand rapidly. Each may have its own distinctive agendas and priorities about which we are completely ignorant. Landis also assumed that violent clashes and invasions of the *Star Wars* variety are exceedingly unlikely. That assumption is of course contestable. A technologically superior community may have no scruples about displacing an inferior one, in much the same way that Europeans displaced Native Americans and Australians from their lands. But if one rules out interstellar Ghengis Khans (or Fermi's paradox comes back to bite us), then some interesting results flow from Landis's computations. It turns out that the pattern of dispersal depends sensitively on the actual strength of expansionary zeal. If motivation falls below a certain critical value, renewed colonization starts to sputter and eventually runs out of steam. In that case, the final configuration consists of compact clusters of colonies surrounded by large unoccupied territory. Above the critical threshold, this marbled pattern gives way to a more pervasive demography. The expansion stops only when the galaxy becomes saturated with colonists, but even then some small patches remain untouched. At the critical value, the final state assumes a so-called fractal structure, with both colonized and uncolonized regions apparent on all scales of size (see Fig. 10).

One unrealistic aspect of Landis's analysis was any element of competitiveness. Recently Robin Hanson redressed this shortcoming by adapting an economic model to the problem of galactic colonization dynamics. The basis of the model is that competition inevitably shapes the pattern of growth. Hanson points out that whatever the motives a community may have for spreading, and whatever the parameters such as travel speed, length of sojourn at new colonies, order of priorities and level of incentive to continue, there will always be a *fastest* wave of migration. Given a sufficiently rich plethora of diverse cultures vying for planetary pastures new, the leading edge of this wave will be determined purely by competitive selection effects. The wave will spread out from the source community to invade nearby territory (which may

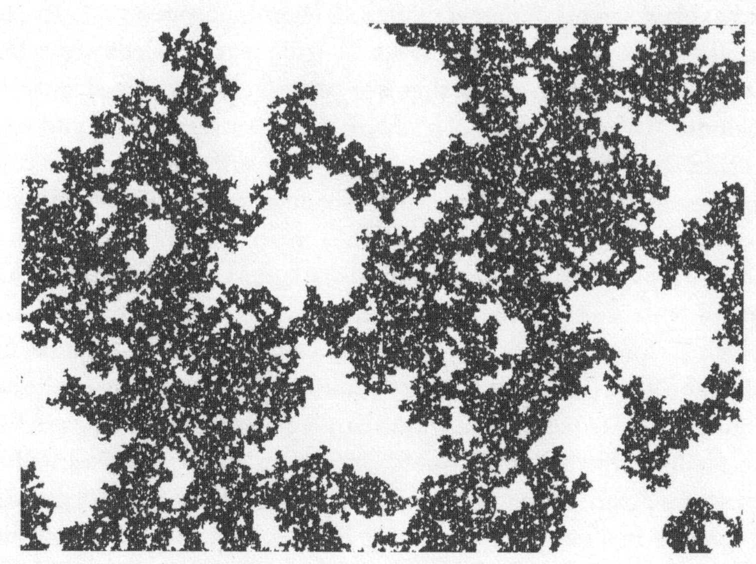

Fig. 10. Computer-generated fractal structure, based on percolation theory. The filled areas represent colonization sites. Note the existence of voids (unoccupied territory) on all length scales.

already be occupied by other, less advanced or less expansionary civilizations), and move on. That is, the wave will move on. Individuals or communities may stay behind, and secondary, slower waves may follow the first, even as the frontier expands apace. In this respect, the migration wave is more akin to a fashion wave than a stampede. If an extraterrestrial community chooses to embark on such a project of expansion, and has the technology and resources to do so, it's hard to see what would stop it, apart from the colonists running into another community doing the same, since there is (presumably) no writ that runs galaxy-wide. The fastest frontier wave is still of course limited by the speed of light, but there is no scientific impediment (as opposed to practical engineering obstacles) to approaching that limiting speed arbitrarily closely.

Hanson finds from his mathematical model that life at the frontier is tough, as indeed it was in the American Wild West. Rapid growth at colonization 'oases' is matched by rapid death between oases: on average, only one 'seed' sent out from an oasis survives to create the next oasis. The 'seeds' here might be, for example, space arks with live colonists, von Neumann machines or small probes with cells to be incubated on arrival. Whatever they are, Hanson draws a stark conclusion: it's all down to staying power. 'A trillion plain seeds are worth as much as a million seeds that are twice as penetrating,' Hanson concludes.[10] There will be a trade-off between seed speed and seed survival; for example, a high-speed seed may suffer more devastating impacts with space dust than a slower competitor. Curiously, colonies with high growth rates fare better if they wait longer before launching new seeds. By contrast, in economically stagnating colonies there will be more pressure to move on and 'ride the wave', because that is where the rich pickings are (whatever those pickings might be – the beauty of Hanson's model is that it doesn't matter). Thus there may be fewer stragglers left far behind the wave than we might intuitively imagine. As a result of the interstellar gold rush, some potential oases will be bypassed – again, rather more than we might expect by analogy with the human experience of terrestrial colonization, but in conformity with Landis's analysis. Our solar system might be located in one such bypassed oasis, which provides another possible resolution of Fermi's paradox.

If the alien migrants were biological organisms rather than machines, there may be a more specific reason why our planet was shunned. Earth has hosted life since very early in its history, so there is a high chance that if ET stopped by, our planet was already seething with micro-organisms, and possibly macro-organisms too. In science fiction, when humans step out of a spaceship on to a verdant planet, they simply take up residence as if it's a duplicate Earth. But this is ridiculous. The chances of alien biology matching the terrestrial variety are very low indeed. Even if DNA is the only viable genetic molecule, there is no reason why the same amino acids in similar combinations would be used as enzymes by all life. Alien and terrestrial life forms simply wouldn't mesh, so the aliens couldn't eat our plants and animals. (So much for the lowbrow science fiction plot that the aliens want us as a source of food.) Conversely, aliens would be unlikely to succumb to terrestrial germs (as they did in H. G. Wells's novel *The War of the Worlds*). Rather than offering an incentive to invade, the biosphere could actually be an inconvenience to the aliens, apart perhaps from the oxygen it has released into the atmosphere. Successful colonization of Earth would probably entail building huge and expensive artificial habitats, or eliminating the indigenous biosphere altogether and replacing it with an alien one – terraforming Earth itself in fact. So contrary to popular lore, our planet's rich and entrenched biology could explain why ET is *not* here.[11]

Absent from Hanson's computations are less savoury scenarios: for example, that uncooperative colonists may be forcibly exiled in seeds shot into the galactic badlands, or ejected from a colony against their will in the space age equivalent of walking the plank. These rejects may roam the galaxy as 'pirates' or skulk unobtrusively in astronomical backwaters. Worse still, they might mutate and evolve into wantonly destructive killers that run amok through space, wreaking havoc – entities known to sci-fi aficionados as 'berserkers'. The application of game theory to such 'good guy, bad guy' competition in a galactic context might yield interesting variations on the simple percolation theory results.

DID THE WAVE PASS THIS WAY?

If an alien colonization/exploration front swept through our region of the galaxy long, long ago, would it have left any traces? Obviously if there was an expanding wave, the aliens (who may have been biological organisms, machines, hybrids, mixtures, or some other entities entirely – see p. 161) will by definition be seeking to achieve something – precisely what, we cannot know. Whatever it is, if it exists in finite quantity (which must be so, or the aliens could get all they need at home), then this Desired Thing will eventually become exhausted, at which point the colony might very well be abandoned. The wave front itself will by then have long moved on. We have no idea when the wave may have passed; it could, for example, have been before the solar system formed 4.5 billion years ago. In this subject it pays to think on astronomical, not human, timescales, and that means anything from 10 million to billions of years. Why? Well, the technical way of expressing it is that we don't know the probability distribution for alien visitation as a function of time, so a reasonable first approximation is to assume it is uniform. What this jargon means is that, in the absence of any good reason to the contrary, there is nothing special about the present epoch, so there is the same chance that aliens will arrive in our part of the galaxy in, say, the next 1,000 years as in any other thousand-year window over a multi-billion-year range of galactic history.[12] So if aliens did visit, it would in all probability have been a *very* long time ago. Clearly the chances of them stopping by within the last few thousand years, and leaving bottles, wires and plastic cups for us to find, are infinitesimal.

Suppose instead that a relatively slow-moving wave passed through our region long ago; it may still be out there somewhere, spreading across the galaxy tens of thousands of light years away. Could we see the leading edge of the wave from Earth? We might, but it's not clear what to look for. *Any* sort of anomaly or physical discontinuity with the shape of a wall would be a good candidate. To take a simple but probably silly example, suppose the frontier colonists power their activities using nuclear fission, and dispose of the waste (very effectively) by dumping it into the host star. There would then be a trail of short-lived radioisotopes in stars close to the moving front, with an abrupt jump ahead of the leading edge, and a systematically declining intensity to the rear (on account of the finite half-lives of the radioactive nuclei). This distinctive pattern would show up in the spectra of the stars in that region of the galaxy. Another (equally speculative) possibility is that the aliens might harvest material from high-mass stars before they blow up, thereby forestalling their demise. If so, supernovae would be distributed irregularly across the galaxy, suppressed for no apparent reason in some regions, and normal in others. If this pattern showed up in combination with weird spectra from stars behind the leading edge, it could be evidence for alien tampering. Unfortunately supernovae are so rare that it may take several millennia to build up the necessary statistical evidence.[13]

Rather than looking for the edge, we could hunt for evidence that the wave had passed through, or near, the solar system in the past. Perhaps the aliens took something that should be here, or left something that shouldn't. In blunt terms, that translates into 'They plundered commodity *X*, and dumped commodity *Y*.' Humans have left many derelict and polluted industrial sites, stripped of raw materials and abandoned as wastelands. Might we identify an alien *X* and *Y*?

There are no obvious signs of ancient industrial activity on Earth itself: no 10-million-year-old mines or quarries or scrapyards. Of course, the scars of industry wouldn't last long on our planet,[14] so it's not clear how conspicuous such evidence might be, or how distinctively artificial it would appear. If we found a triangular crater, for example, even though it was now buried, it would be striking evidence of artificiality. Geologists have discovered hundreds of craters, both on the surface of Earth and buried, but so far they are all approximately round, that being the natural shape created by both cosmic impacts and volcanic eruptions. There *is* a weird geological anomaly in Gabon, Africa, known as the Oklo natural nuclear reactor. It is a substantial rock formation with an unusually high uranium content that apparently 'went critical' about 2 billion years ago, creating a self-sustaining chain reaction and generating a lot of heat and radiation in the process, the products of which are detectable today. Oklo is certainly an unusual geological relic, although invoking alien nuclear engineering is a bit of a stretch. It does, however, illustrate the sort of anomaly we might watch out for.

Plutonium offers a more promising possibility. This radioactive element is manufactured in nuclear reactions, and is present in the waste from nuclear power plants and in the fallout from nuclear explosions. It will remain in the environment in declining concentration for millions of years. If we ever found an ancient plutonium deposit (on Earth, or anywhere else in the solar system), it would constitute strong evidence for alien nuclear technology.[15] Using radioactive dating we could even work out when the nuclear engineering took place. Another potentially suspicious geological feature would be a mineral deposit of peculiar size, shape, location or composition that might point to an ancient waste dump, especially if buried in an 'unnatural' setting. All these suggestions are extremely far-out guesses of course, but the point I want to make is that nobody (as far as I know) has made a systematic search of geological records for anomalies that might hint at alien tampering.

Away from Earth, the possibilities multiply. Moons, comets and asteroids would all provide an ideal source of raw materials for alien technology with the added attraction of being located in low surface gravity environments. Precision-formed tunnels or bridges on one of those bodies would be a dead giveaway. Less dramatic oddities might provide evidence for mining activity, such as spoil heaps or (again) odd-shaped craters. Amazingly, Eros, one of the first asteroids to be studied in detail, has some square craters! The spacecraft NEAR Shoemaker photographed them in 2000. There is a natural explanation in this case, though. Straight fault lines are common geological features, and where they intersect approximately at right angles, a roughly square depression can form. A better bet would be to look for spiral craters, of the kind that might be made by open-cast mining when a vehicle goes round and round. On Earth, spiral craters would soon erode to appear round, but on an asteroid or on the Moon the spiral form would survive for much longer.

A more subtle signature of mining, or resource-harvesting, could be left in the chemistry and morphology of the debris. For example, if nuclear explosives were used to blow an asteroid to bits, the fragments might carry evidence in the form of distinctively fused surfaces, like the piece of trinitite I have, salvaged from the first atomic bomb test at Alamogordo in New Mexico. If a meteorite were ever discovered with traces of unusual radioactive isotopes, that could also constitute evidence for the rock having been blasted by a nuclear explosion.

ONE OF OUR PLANETS IS MISSING

Let me now focus on scenario *X* – the anomalous absence of something. How about this: aliens passed through our part of the galaxy a long time ago harvesting comets for their water and organic material? It is a plausible enough strategy, one in fact being considered by our own space futurologists. A comet's water can be electrolysed and the hydrogen used for a nuclear fusion reactor. As a bonus, comets are enriched in deuterium – heavy hydrogen – an especially good nuclear fusion fuel. The hydrocarbons that make up part of the dirt of the 'dirty snowball', as comets are often described, can be used to produce a range of synthetic materials, and as a food source. Most comets are believed to originate in the so-called Oort cloud (after Jan Oort, the astronomer who proposed the idea), which consists of a trillion small icy bodies located about a light year from the sun. It is likely that other stars have their own comet clouds at similar distances. Because these far-flung 'dormant' comets are only loosely bound to their parent stars, they would make ideal sources of raw material for interstellar travel, obviating the need for a spacecraft to enter the deep gravity well of the star and then climb back out.

From time to time gravitational disruption sends one of the comets from the Oort cloud plunging sunward on an elongated elliptical trajectory, whereupon it blazes in the night sky in familiar spectacular fashion. But there is also a good chance that the gravitational disturbance will kick a comet the other way – propelling it into interstellar space. If the solar system is typical, and other stars have comet clouds too, then the comets ejected from them should sometimes come our way and enter the solar system. If an extra-solar comet paid us a visit, it would be seen travelling on a hyperbolic rather than elliptical orbit, i.e. moving too fast to be from the Oort cloud. So far no such comet has been seen, which is a bit puzzling. Perhaps our neighbour stars are light on comets for some reason. Did ET steal them all? If future astronomical searches reveal a *systematic* depletion of comets in some star systems but not others, it could suggest harvesting. Similarly, if a population of comets strongly depleted in deuterium is found (something that can be determined from the comet's spectrum) it might hint at them being mined for nuclear fuel.

Could an alien technology commandeer entire planets and pull them apart for raw material? There is a range of masses from comets up through icy planetesimals, minor planets like Pluto and moons like Titan, to terrestrial and giant planets. If ET can hijack comets, why not one of these larger bodies? The Princeton physicist and futurologist Freeman Dyson speculated on this possibility with his proposal for 'Dyson spheres' (more on that soon). But how do you pull a planet apart? It's certainly not easy. The total energy needed to blast Earth to smithereens, for example, is equivalent to the total power output of the sun for several days. Slamming another planet into it wouldn't work – in fact, that's what already happened when the proto-Earth was struck by a Mars-sized body about 4.5 billion years ago. The outer layer was stripped off (and became the Moon), but the rest of the material merged to make a *bigger* planet. A neat idea for disassembling planets was put forward by the writer Greg Bear in his apocalyptic science fiction novel *The Forge of God*.[16] Bear tells the story of an alien civilization releasing self-replicating von Neumann machines that run amok, sweeping through the galaxy, ripping planets to bits. The clever trick the soulless plunderers use is to drop a massive slug of 'neutronium' (a hypothetical ball of neutrons possessing nuclear density) into Earth, followed by an equivalent mass of antineutronium (its antimatter counterpart). The two slugs spiral in together towards the Earth's core, where they eventually annihilate each other, releasing enough energy to blow the planet apart and hurl its hapless inhabitants into space.

All of which brings me to a persistent space age story, which is that the asteroid belt between Mars and Jupiter might be the remnants of a planet that somehow got itself blown up. It's true that there is a curious 'gap' there where a planet might have been, but the total mass of the asteroids isn't enough to constitute an entire planet. The conventional explanation is that most of the debris in this region of the solar system was drawn away by the powerful gravitational pull of Jupiter, thus preventing a planet forming, but we could speculate that an ancient super-technology pulled the planet apart, took whatever it needed, and then moved on, leaving the rubble to form the asteroid belt.

Rather than go to the trouble of rending already-formed planets asunder, rapacious aliens might find it easier to simply intercede before the planets fully aggregate in the first place and make off with all the good stuff, leaving the dross. Evidence for such selective harvesting could be obtained from the discovery of planetary systems with anomalous chemical and/or physical composition. At this stage, astronomers do not have a sufficient understanding of the process of planet formation to identify such anomalies, but with the increasing tally of extra-solar planets being discovered, that shortcoming should soon be rectified. A number of star systems are known in which the process of planet formation is under way at this time; they would be a good place to look for signs of large-scale alien astro-engineering.

In principle, it would be possible for a super-technology to carry off an entire intact planet by manipulating the chaotic nature of some planetary orbits. Beginning with a nuclear explosion to deflect a small asteroid and bring it into collision with a larger body, a series of carefully controlled manoeuvres could have an accumulating and amplifying gravitational effect over an extended period. Eventually a planet's orbit could be destabilized enough for it to be flung out of the planetary system altogether. Subsequent encounters with other stars would provide the opportunity for additional gravitational slingshot boosts to increase speed. The hijacked planet could then be used as a handy space ark for traversing the galaxy, an idea foreshadowed by Olaf Stapledon in his 1937 science fiction classic *Star Maker*.[17]

ABSENT EXOTICA

Planets are not the only things that could go missing. Theoretical physicists are masters at predicting things that might exist, but don't seem to be there. Exotic subatomic particles with whimsical names such as neutralinos, shadow matter and axions grace the theorists' lexicon, but haven't yet shown up in the lab. At the other end of the mass range are mini-black holes, quark stars and cosmic texture, to name but a few. Did ET make off with them? Clearly, extreme caution is needed before considering alien culpability. Remember Bayes' rule: the hypothesis that aliens are the correct explanation for the anomalous absence of something is only as good as the prior probability of an alien super-civilization in the first place. That may be very low. By contrast, the prior probability that Professor A's theory of the so-and-so particle, or Dr B's prediction of such-and-such an astronomical object, is simply wrong could be a lot higher.

Some of the 'missing' particles may yet show up; they may, for example, constitute the famous dark matter that pervades the cosmos but has yet to be identified. It is also possible that the theorists got carried away. Set against this, some unconfirmed predictions are fairly robust. A good case in point concerns particles known as magnetic monopoles, about which some explanation is in order. Familiar magnets always come as 'dipoles', with a north pole at one end and a south pole at the other. A magnetic monopole, if it exists, will be an isolated *N* or *S*. You can't make a magnetic monopole by chopping a bar magnet in two; you just make two dipoles, with a new *N* and *S* respectively appearing on opposite sides of the cut. But physics has a neat place in its mathematical closet just waiting for magnetic monopoles to fill it. After all, electric charges come as monopoles (+ and –), and electromagnetism is otherwise completely symmetric between electricity and magnetism. The British physicist Paul Dirac developed a theory of magnetic monopoles in the 1930s, and even figured out what their magnetic 'charge' should be. Then in the 1970s theoretical physicists rediscovered the concept of magnetic monopoles while attempting to formulate a unified description of electromagnetism and the two nuclear forces, theories known collectively by the pithy acronym of GUTs (for 'grand unified theories'). Direct searches for magnetic monopoles have been made over the years, by scouring iron deposits, the sea floor, cosmic rays, and even Moon rock. No luck. There was a memorable false alarm in 1982 when a Stanford University physicist, Blas Cabrera, thought he'd found a monopole using a clever technique. Cabrera had a wire ring that he made superconducting by cooling it to near absolute zero. If a magnetic monopole by chance passes through the hole in the middle of the ring, it will abruptly generate an electric current. What's more, Dirac's theory tells us exactly how much this current should be, and that's the value Cabrera claimed he saw. Alas, his results were not confirmed, and were dismissed as a glitch in the equipment.

A distinctive feature of GUT magnetic monopoles is their huge mass, predicted to be a thousand trillion times greater than a proton, making them heavier than a bacterium. With a mass like that, it's no wonder they haven't been made in the lab – the energy requirements are stupendous. But what about in the big bang that gave birth to the universe 13.7 billion years ago? Plenty of energy to spare there. In the late 1970s cosmologists began to realize that the universe should be bursting with primordial magnetic monopoles made by the searing heat a split second after the universe received its starting orders. Their puzzling absence prompted Alan Guth of MIT to propose a drastic solution. Maybe, said Guth, the universe abruptly leapt in size by a factor of trillions and trillions just after the monopoles got made, thus diluting their density to unobservable levels. He called this explanation of the missing monopoles 'inflation' (to distinguish it from the familiar less frenetic cosmological expansion). It was soon found that inflation explained a lot of other cosmological mysteries too, and today it forms part of the standard model of the early universe. But the inflation theory has been challenged by some cosmologists. Although it has a lot of supporters and there is good observational evidence in its favour, it is far from secure. So the mystery of the missing monopoles hasn't gone away yet.

We can't be sure that the lack of monopoles is universal – maybe it's just our region of the galaxy that is affected. Are the aliens to blame? Why would magnetic monopoles be of use to them? Well, it turns out that monopoles would be *the* power source of choice for any self-respecting super-civilization. That's because an N and an S are not just oppositely charged, magnetically speaking. They are also antiparticles of each other, which means if they come together they neutralize their magnetism and annihilate, releasing their mass as energy ($E = mc^2$ again). You could have a jar of norths on one side of your lab and a jar of souths on the other side, and when you are ready just mix them together and . . . poof! The blast would be some billion billion times greater per gram of material than thermonuclear fusion (as employed in hydrogen bombs).[18]

If the absence of magnetic monopoles is explained by alien sequestration (rather than inflation), might we see evidence for some of the 'poof' events described above? Well, possibly. The liberated energy would be released in the form of lighter subatomic particles, including the humble electron – and its antimatter opposite number, the positron. Recently, high-energy electrons and positrons have been detected coming from space, using an instrument slung beneath a balloon and flown 37 kilometres (23 miles) above Antarctica.[19] The origin of these particles has caused a certain amount of head-scratching among astrophysicists. They might be coming from a hitherto overlooked pulsar, or from something more obscure, such as the annihilation of dark matter. As yet, nobody has suggested exhaust from a monopole-powered alien factory. . . .

Another example of a longstanding theoretical prediction, as yet unverified, is the so-called cosmic string – an ultra-thin tube packed full of energy at such concentration that a mere kilometre length would outweigh the Moon. As with magnetic monopoles, cosmic strings might have been made in the big bang. They are so heavy their gravity would bend light rays from distant galaxies, creating distinctive double images. From time to time astronomers claim to have discovered cosmic strings, but then the evidence goes away; whether or not they really exist remains an open question. A cosmic string would pack even more punch than a pair of magnetic monopoles. In effect, the string is a nanotube that traps the colossal primordial energy the universe had at a trillion trillion trillionth of a second after the big bang. If that energy could somehow be extracted in a controlled way – for example, by shrinking a closed loop of string to zero size – the aliens wouldn't need to worry about their electricity bills for a long time. Cosmic strings are taken seriously by many physicists and cosmologists,[20] and their apparent absence is a source of disappointment, if not outright puzzlement, to some. Magnetic monopoles are more firmly established by theory than cosmic strings (although they originate from similar concepts), so their peculiar absence is more demanding of an explanation.

In this chapter I have restricted the discussion to galactic exploration and colonization, but a sufficiently advanced and motivated technological civilization could spread to neighbouring galaxies, and ultimately across the entire observable cosmos. Even if the universe we observe at this time has not been 'taken over' by one or more super-civilizations, there is plenty of time in the future for it to happen. And who knows, maybe our own descendants will be part of this glorious cosmic adventure.

7
Alien Magic

Any sufficiently advanced technology would be indistinguishable from magic.

Arthur C. Clarke

SIGNATURES OF DISTANT SUPER-TECHNOLOGY

If we were to encounter alien technology far superior to our own, would we even realize what it was? Think how a laser or a radio would seem to a tribe of rainforest dwellers who have never been in contact with the outside world. Now imagine a technology a *million* or more years in advance of ours: it might well appear miraculous to us. All of which presents new SETI with a serious problem. How can we look for signatures of alien technology when we have no idea how it would be manifested? In the previous chapter I suggested some ways in which an advanced civilization spreading across the galaxy might leave traces of its activity. But all the examples I gave were based on extrapolations of twenty-first-century human physics, and so are tainted by anthropocentrism. Suppose that alien technology is based on principles that are completely beyond the ken of our best scientists?

One way to tackle the problem is to consider very general physical effects – effects that might be expected even from 'magical' technology. In 1964 the Russian astronomer Nikolai Kardashev proposed a measure of alien technological advancement based simply on energy consumption. Now it's true that this Soviet-era heavy-industry criterion is yet another example of SETI parochialism. Today we might attach more significance to terabytes than megawatts; tomorrow, who knows? However, there is a good reason to stay with Kardashev's classification scheme when considering alien technology that might be very distant from Earth. Given the current limitations of our instruments, we would probably be able to detect alien industry only if it produced a very large energy footprint.

Kardashev defined a Type I civilization as one that uses all the energy resources of its home planet to power its industry. A Type II civilization is one that requires the total energy output of its parent star, while a Type III civilization would need a whole galaxy to run its projects. To this we might add Type IV: a civilization that commandeers the entire cosmos. To date, there is no evidence for Kardashev civilizations of any numerical status, although Type I would be hard to spot. Type II is an interesting case, because utilizing the total power output of a star – no mean feat – would definitely leave tell-tale signs. One way a civilization might accomplish it was suggested in 1959 by Freeman Dyson.[1] Inspired by Stapledon's novel *Star Maker*, Dyson envisaged the construction around a star of a spherical shell of matter with a radius similar to that of a planet's orbit, made from a dense swarm of particles designed to collect all the star's heat and light for as long as it keeps shining. Compare this energy bonanza to the paltry one billionth of the sun's output intercepted by the Earth. The construction material would come from planets and asteroids, after pulling them apart to build the necessary structures. The construction would, of course, be a gargantuan undertaking, but it's theoretically possible. A Dyson sphere would dramatically alter the light spectrum of the entombed star, creating a noticeable infrared glow that could be identified by inquisitive astronomers, even on the far side of the galaxy. Searches for Dyson spheres have actually been made, by analysing the database of the Infrared Astronomical Satellite (IRAS), so far without success.[2]

A Type II civilization capable of reconfiguring a planetary system might consider a more attractive option, first mooted by John Wheeler, the physicist who coined the term *black hole*. Wheeler envisaged building a shell of matter around a spinning black hole, a strategy offering distinct advantages over Dyson spheres. First, black holes don't inconveniently burn out after a few billion years (they are, after all, the remnants of stars that have already burned out). Second, they are ideal dumps for unwanted rubbish: anything that falls into a black hole is irreversibly swallowed and permanently obliterated. Third, they can be used to launch spacecraft at a significant fraction of the speed of light (see below). Finally, a black hole can release far more energy than a star ever can through nuclear fusion. The secret of a black hole's prodigious power lies with its rotation. All stars spin, and when the core of a star collapses to form a black hole the spin dramatically increases, a result of the law of conservation of angular momentum. Young neutron stars, which are black hole near misses, have been observed spinning as fast as hundreds of revolutions per second. A spinning body contains more energy than a static one, and because energy and mass are equivalent, one may express the energy of rotation as a fraction of the total mass. In the case of a black hole, up to 29 per cent of the total mass can be in the form of rotational energy, and in theory this entire fraction can be extracted and used. Compare 29 per cent with the miserable 1 per cent of its mass that a star typically radiates as heat and light accumulated over its multi-billion-year lifetime. Obviously, spinning black holes represent an energy cornucopia. If raw power is what you want, black holes are it.

Based on calculations by Roger Penrose, Wheeler dreamed up the amusing scenario depicted in Fig. 11, in which trucks containing industrial waste are dropped on a carefully calculated trajectory towards the spinning black hole. When they enter a region close to the surface of the hole (known technically as the ergosphere), a remarkable transformation becomes possible. The trucks spill out their contents in such a way that the waste is devoured by the black hole. For certain trajectories, the empty trucks get propelled away from the ergosphere at high speed, zooming off with more mass-energy than the laden trucks originally had going in. Ultimately the additional energy has to come from somewhere, and in fact it comes from the rotational energy of the hole; every time the trick with the trucks is performed, the black hole's angular speed drops a bit. The good times will not last for ever – eventually all the rotational energy will be extracted and the civilization will be obliged to decamp elsewhere. But at present human levels of energy consumption, a black hole could meet our energy needs for at least a trillion trillion years. To the best

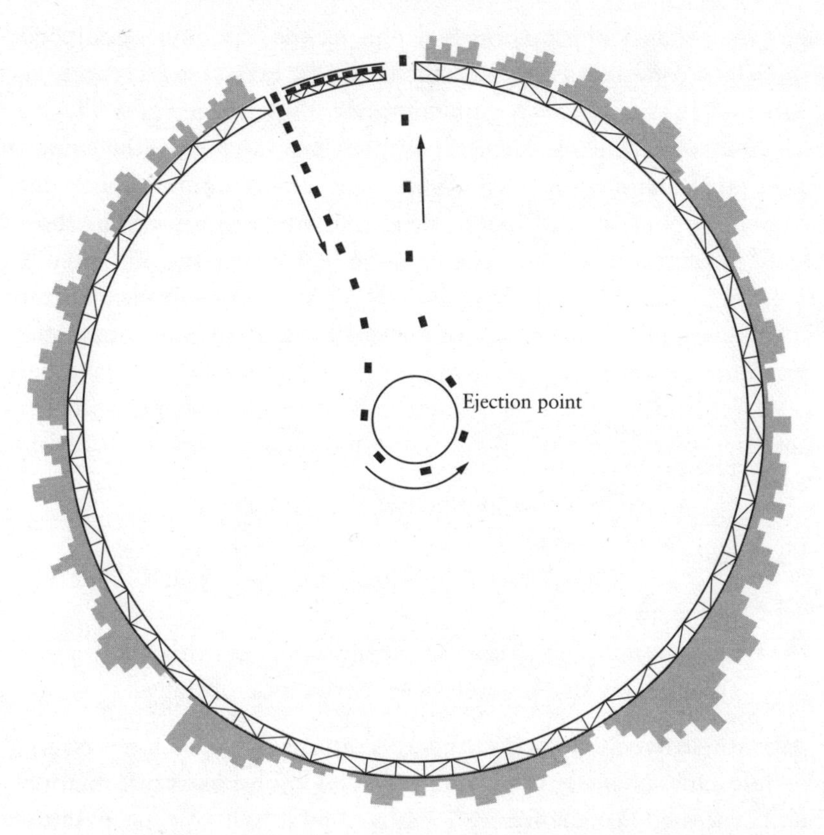

Fig. 11. Whimsical depiction of energy extraction from a rotating black hole.

of my knowledge, no SETI searches have targeted black holes, perhaps because they are hard to detect.

TECHNOLOGY AS 'NATURE-PLUS'

To go beyond crude identifiers of alien technological activity, such as energy and resource usage, leaves us groping for a familiar reference point, with the inevitable temptation to fall back on human experience. Even science fiction tends to portray alien engineering as closely analogous to our own. In the 1980 movie *Hangar 18*, for example, a flying saucer is investigated by the simple expedient of pressing a few buttons to see what happens. The giant spacecraft in *Independence Day*, despite being the product of a million-year-plus technology, comes equipped with 1990s computer consoles, sans firewalls. Even in more carefully crafted science fiction, alien artifacts appear recognizably as *machines*, in the twentieth-century understanding of the term: regular in geometrical shape, made of metal or some superior substitute, often inert except in response to a deliberate prod, and built on an everyday scale of size. But advanced alien technology might be nothing like that at all. In fact, in contemplating the activities of a super-intelligence it pays to clear your mind of all preconceptions. To help this process, consider a hypothetical alien technology that:

- Is not made of matter.
- Has no fixed size or shape.
- Has no well-defined boundaries or topology.
- Is dynamical on all scales of space and time.
- Or, conversely, does not appear to do anything at all that we can discern.
- Does not consist of discrete, separate things; rather it is a system, or a subtle higher-level correlation of things.

We are so wedded to the human concept of a machine as, for example, chunks of metal with buttons and knobs, or as information being processed (as in software), that we find it hard to conceptualize technology involving levels of manipulation above these. What do I mean by this? A conventional machine such as a car moves matter around in an organized way. Information technology on the other hand moves *information* around in an organized way. For example, Photoshop on my computer can rotate an image. When that happens, matter moves too, namely electrons in the computer's circuitry, but we wouldn't recognize the technology in action by observing the electrons – we see it via the complete image.

One way to think about information is as a 'higher level' concept than matter. The higher level builds on, but transcends, the lower level. Thus software – an abstract concept – invariably requires physical hardware to support it: swirling bits of information inside a computer, or sense data in the brain, need switches or neurons. Now, I ask, are these two conceptual levels – matter and information – all there is? Five hundred years ago the very concept of a device manipulating information, or software, would have been incomprehensible. Might there be a still *higher* level, as yet outside all human experience, that organizes information in the same way that information-processing organizes electrons? If so, this 'third level' would never be manifest through observations made at the informational level, still less the matter level. There is no vocabulary to describe the third level, but that doesn't mean it is non-existent, and we need to be open to the possibility that alien technology may operate at the third level, or maybe the fourth, fifth . . . levels.

To think creatively on this topic, we must even be wary of notions like 'control' and 'manipulation' and 'design', for these are also human categories that may turn out to be short-lived. The arbitrary separation of objects into 'natural' and 'artificial' is something that we take for granted, but as I shall argue in the next chapter, it is a purely cultural distinction. Technology is, in the broadest sense, mind or intelligence or purpose blending with nature. Importantly, technological devices don't subjugate nature; the devices still obey the laws of physics. Technology *harnesses* the laws; it does not override them. So to say that a radio or a laser or an obelisk on the Moon is 'unnatural' doesn't mean it isn't part of nature. The best way I can think to express it is to say that technology is nature-plus. (Art is also nature-plus.) The value that is added by technology is a very specific amalgam of constraint and liberation, most obviously associated with purposeful goals. A washing machine can't bake bread, but it can do what unmodified nature can't, namely, wash and rinse and spin-dry clothes, which is what it is designed to do. A computer can't fly, but it can prove the four-colour theorem, which is not on Mother Nature's agenda, anywhere, as far as I know. However – and this is the key point I want to make – technology of *that* sort – our sort – may be only *one* way that nature becomes nature-plus. And we may utterly fail to recognize or appreciate the significance of a more sophisticated form of nature-plus, even if it were staring us in the face.

A machine is characterized by possessing a certain relationship between the parts and the whole: the components cooperate in an systematic way to fulfil a global function. William Paley famously drew an analogy between a watch and a living organism, noting that both consist of a coherent overall system of mutually supportive parts,[3] a concordance that is today explained by Darwinian evolution. But machine and biological functionality represent only one way that parts and wholes might interrelate in a special and unusual manner. In fact, we know of another example already: quantum systems. Quantum mechanics is the crowning achievement of twentieth-century physics, and its successful predictions and explanations range from particle and nuclear physics to cosmology – and much in between. Quantum mechanical principles underlie the laser, the transistor, superconducting magnets and many other items of human technology. The theory explains nearly everything from the big bang to nuclear power to chemistry to electricity. So we have to take its predictions seriously.

One prediction made by quantum mechanics is that a part is properly defined only in relation to the state of the whole of which it is in turn a part. This Zen-like description can best be understood with an example. An atom can behave either as a wave or as a particle. In isolation it is neither of these specifically; its status is undecided. But placed in the context of a larger system, its inherently ambiguous nature may be resolved. Here is how. We can construct a type of microscope that will determine the position of a particular atom, call it *A*. After the measurement, *A* will be 'an atom-at-a-place'. Alternatively, we can construct an apparatus that will bring out the wave-like nature of the atom, in which case *A* will then be 'an atom-with-a-speed' (a quantum wave describes the atom as having a specific momentum). The crucial point is that, according to quantum theory, *A* cannot be *both* 'at a well-defined place' and 'possessing a well-defined speed' *at the same time*. Which aspect of *A*'s dual identity is manifested, wave or particle, depends on which type of apparatus *A* interacts with; that is, on the arrangement of the whole environment. Now the system 'atom *A* plus apparatus' is itself a collection of atoms, so the particular configuration and state of all the atoms taken together serves to define the nature of the individual atom *A*. And this is true in general: all atoms that interact with larger systems are defined in part by the totality of atoms, while in turn that totality is made up of the parts. There have been many attempts to capture this 'up-and-down' whole–part interdependence of quantum systems. Niels Bohr likened it to yin and yang. David Bohm described it as 'implicate order'.[4] In recent years it has been dubbed 'quantum weirdness'.

Quantum weirdness, living organisms, minds and designed machines all provide examples in which wholes and parts interrelate in different ways. It would be naïve to suppose that the foregoing list is exhaustive. There could be many ways that whole–part relationships could differ from anything in our experience. After all, a hundred years ago, who would have suspected that atoms behave like *that*? Truly advanced alien technology might manifest itself by an entirely new form of whole–part interrelationship. And just as quantum weirdness is uncovered only with very special apparatus, so alien technology might go unobserved and unsuspected, because we are not viewing it with the equivalent of . . . well, a Bose–Einstein condensate beam-splitting interferometer.

FANTASTIC SUPER-SCIENCE

New SETI demands an uneasy compromise between the need to think about alien technology as creatively and imaginatively as possible, while at the same time taking care not to stray across the sometimes blurred line between legitimate science and science fiction. Science fiction writers are generally happy to play fast and loose with the laws of physics, mingling science, speculation informed by science, and outright fantasy. That's okay: they have literary licence on their side. But a scientific appraisal of SETI needs to do better.

Take that old bugbear of space travel – the finite speed of light – which has stood in the way of many a good sci-fi drama. As I have explained, Einstein's theory of relativity forbids anything from breaking the light barrier, so if we understand the laws of physics correctly, neither spacecraft nor messages can go faster than light. The distances between stars are measured in light *years* (the distance light travels in a year), which means interstellar travel is completely unrealistic in a human lifetime, unless speeds approaching that of light are attainable. Even then, there are problems. At, say, half the speed of light, a spacecraft would face numerous hazards, such as impact with micrometeorites that would explode like bombs on its surface. Such complications may turn out to be so daunting that interstellar high-speed travel is for ever unattainable in practice. However, it's also possible that an advanced technological community will eventually solve the practical problems: for example, by detecting oncoming micrometeorites and zapping them with a laser before impact. So while it may or may not be a realistic proposition, travel at close to the speed of light is legitimate speculation because it does not conflict with basic physics. But travel faster than light is not.

Another way to cross space quickly, and much beloved of science fiction, is teleportation. You just scan something – a human being say – and 'beam' the information to the destination, where the object is reconstructed. This trick is performed in *Star Trek* as a cheap way to get the astronauts down to planetary surfaces and up again (it also speeds the story line along). Is teleportation valid science? Well, up to a point. So long as the beaming doesn't happen faster than light, some sort of information transfer might be possible. As a matter of fact, physicists have already achieved a limited sort of teleportation, in which information about the state of a quantum particle is beamed between field stations using lasers. But, as pointed out by Lawrence Krauss in his book *The Physics of Star Trek*, there are fundamental reasons why scanning every atom in your body and reassembling the whole thing at the other end would involve overwhelming technological obstacles.[5] For a start, storing the total information content of a body scan would require a stack of disks that would reach a third of the way to the centre of the galaxy. Not physically impossible, maybe, but probably too expensive even for a galactic super-civilization. Too bad, Scotty.

In *Contact*, Carl Sagan proposes a wormhole as a way of moving his heroes through space in next to no time. Wormholes, which are loosely like stargates, are also a popular proposal for time travel (see p. 121). They don't seem to violate any laws of physics so far known, but the existence of a wormhole would require prodigious quantities of a type of exotic matter known to exist only in ultra-microscopic quantities.[6] Unless we discover a new source of this exotic matter, then large traversable wormholes will probably remain for ever fictional.[7]

Readers who think I am being a party pooper should take heart. Even if we remain constrained by the accepted laws of physics, it's still possible to conceive of all manner of mind-bending scenarios. How about techno-savvy alien engineers taking up residence inside hollowed-out worlds or ring-shaped tubes? Or hive societies composed of tangled magnetic threads constructing complex plasma patterns spanning interstellar space, like cosmic termite mounds made of ionized gas? Or beings made from pure gravitational energy that reconfigure spacetime into weird shapes? Such feats of astro-engineering don't *seem* to violate any laws. (It's always hard to know for sure. There may be hidden assumptions that, on closer inspection, fall foul of some law.) That doesn't mean they are going to become a reality, of course. The aliens might not be interested, or may be prevented by political or financial or even ethical considerations from embarking on ambitious projects of this sort. But we can still contemplate these fantastic undertakings, and wonder if they would present a signature detectable from Earth.

FLAWS IN THE LAWS

The examples I discussed in the previous section fall into the category of speculations which, on the surface, appear to conform to our best understanding of science, but may present such formidable practical challenges that they may never be implemented. Pushing the boundaries of legitimate physics that far inevitably comes up against the question of whether twenty-first-century human science is so reliable it can be applied to an alien civilization far in advance of our own. Suppose there are flaws in the laws as we currently understand them? Can we be *absolutely* sure about the speed of light, say?

Now, it's true that there are laws and there are laws. In secondary school children learn Ohm's law of electricity, which says that the current through a resistor rises in proportion to the applied voltage. But Ohm's law is not really a basic law at all; in fact, there are materials never envisaged by Ohm for which it goes wrong. On the other hand, the no-faster-than-light law *is* basic and universal, and may well be for ever non-negotiable. The trouble is, at any given time scientists can only state the laws of physics to the best of their current understanding. Who knows whether a future advance will show one of the cherished laws to fail under certain circumstances? In science, the last word is never said; there is always room for revision in the light of new evidence. All one can claim is that some laws are more deeply entrenched than others.

A case in point is the second law of thermodynamics, which may well be the most fundamental law in the universe. It applies to absolutely everything, no exceptions. Put simply, it says that in closed systems the total entropy (roughly speaking, disorder) can never decrease. Translated into a simple example, the second law forbids heat from flowing spontaneously (that is, without the expenditure of energy) from cold to hot bodies. The British astrophysicist Arthur Eddington once expressed the sacrosanct nature of the second law dramatically:[8] 'If your theory is found to be against the second law of thermodynamics I can give you no hope; there is nothing for it but to collapse in deepest humiliation.' In speculating about alien superscience, then, the second law of thermodynamics should be the last one to go. And that knocks on the head another popular idea: powering a spacecraft by 'mining the quantum vacuum' for energy. Let me explain. When quantum mechanics is applied to the electromagnetic field, in addition to explaining how light and matter interact, the theory predicts something truly remarkable: that a region of space devoid of all matter and all light – indeed, all particles of any sort – will nevertheless still possess some energy. The irreducible energy of empty space is called 'the energy of the quantum vacuum'. And it really exists. You can detect it as a tiny force of attraction between metal surfaces. Astronomers have also measured what looks to be the same thing on a cosmological scale, although they have given it a more mysterious name – 'dark energy'. It's the stuff responsible for making the universe expand faster and faster.[9] Vacuum/dark energy is there all right, with a density of a little less than a joule per cubic kilometre. Could it therefore be 'mined' to power a starship; say, by using a big scoop to harvest the vacuum/dark energy, and then converting it into electricity for a plasma drive? This strategy would eliminate the need for rocket fuel, since in space there is plenty of vacuum available.

Unfortunately the quantum vacuum drive won't work, for the same reason that nineteenth-century perpetual motion machines were all non-starters: they violate the second law of thermodynamics. In the 1800s inventors speculated about powering a ship from the heat of the ocean. After all, seawater contains over half a million joules of heat per litre, merely by having a temperature a few hundred degrees above absolute zero. Can't all that heat energy be used to run a turbine? The answer is yes, but only if there is a sink of heat at a lower temperature than the source. Heat pumps are powered by transferring heat from a hot to a cold reservoir and extracting energy on the way. The point is, there has to be a temperature differential in there somewhere. Similarly with the quantum vacuum: if there is a lower-energy vacuum state into which one can dump the dark energy, then you'd be in business with your interstellar drive. But as far as we know there isn't a lower-energy state, or rather, if there were, nature would already have short-circuited it, with dire consequences for the universe.[10] Conclusion: in the absence of a sink of energy, you can't use the quantum vacuum to power a spacecraft.

Levitation is another popular fictional device. It captured my imagination from the moment I read about Dr Cavor's handy gravity-screening substance 'cavorite' in H. G. Wells's novel *The First Men in the Moon*. Wouldn't it be nice to dispense with all those noisy polluting rockets by simply pressing a button and floating serenely to the stars! Sadly, that proposal is another no-hoper. The snag this time is that cavorite violates a founding tenet of the law of gravitation, which requires all forms of matter and energy to fall equally fast, and in the same direction (i.e. down instead of up). Galileo first discovered this, and Einstein incorporated it into his general theory of relativity as a fundamental principle. Without it, our understanding of space, time, astrophysics and cosmology would fall apart, so scientists are not about to relinquish this principle in a hurry. Theoretically, levitation could be achieved using the same quantum vacuum energy I just discussed, but in practical experiments this energy comes in such tiny amounts it can't overcome the much bigger gravitating effect of matter.[11]

Speculation about alien super-civilizations doing super-science and deploying super-technology is certainly great fun, but it needs to be tempered with a healthy scepticism. There is no doubt that twenty-first-century science is incomplete and provisional, yet it still represents the most reliable approach to knowledge, with a wealth of understanding and experience accumulated over several centuries of careful investigation. In the search for alien intelligence, it is as well to adopt a pragmatic view and go with our current picture of science as the best there is so far on offer to guide us, while being open-minded about the possibility of surprises ahead. The future may well prove some of our basic science to be wrong, but if we take an anything-goes approach to contemplating alien technology, then all we get is speculative anarchy with no useful predictive power. The aliens may be able to travel faster than light, or beam each other across space, or levitate, or (though probably not) make heat flow backwards from cold to hot. But in that case we are off in fantasyland, and we might as well give up thinking about SETI altogether.

8

Post-Biological Intelligence

The machines are gaining ground upon us; day by day we are becoming more subservient to them.

Samuel Butler (1863)[1]

If granted full rights, states will be obligated to provide full social benefits to them including income support, housing and possibly robo-healthcare to fix the machines over time.

Robo-rights, UK Department of Trade and Industry report[2]

CLOSE ENCOUNTERS OF THE ABSURD KIND

Fifty years ago I was a teenager and knew nothing about SETI. My mental image of an alien owed much to the Mekon, leader of the Treens of North Venus, nemesis of the clean-cut hero Dan Dare of the Earth Alliance. At least, that's how *The Eagle* comic depicted it. I assumed that if the flying saucer stories were true, their occupants would, like the Mekon, be humanoids with a large head (implying a big brain), and a shrunken, atrophied body (no longer important). Evidently I was not alone in this belief, because accounts of ufonauts often described them as hairless dwarfs with big heads and large, staring eyes, an image now so entrenched it is a cliché (see Fig. 12). Steven Spielberg reinforced that representation in the movies *Close Encounters of the Third Kind* and *E.T.*, in which the aliens resemble big-brained children.

It's absurd. Many fallacies underlie the popular representation of aliens, and serve to undermine the credibility of the close-encounter reports. The first is to suppose that evolution on another planet would parallel Earth's so closely that intelligent beings would assume a humanoid form. Intelligent aliens might just as well resemble whales or octopuses or giant birds or none of these: they might have a body plan that simply does not exist on Earth and would strike us as utterly bizarre. Another fallacy is a misplaced extrapolation of Darwinian evolution. The usual argument runs thus. If brainpower is what counts and the rest of the body becomes an encumbrance, then natural selection will operate to produce Mekons and ETs. But this reasoning is flawed. Once technology advances to the point where a community can exercise choice over who survives and who doesn't, pure natural selection breaks down. When active genetic modification becomes possible, then the further course of evolution can be determined by design. Whether an alien species would in fact choose to use genetic enhancement to produce bigger brains and smaller bodies is another matter. They may have ethical or other reasons to desist. On Earth, there is strong resistance to the prospect of GM humans, just as there

Fig. 12. Popular image of what an alien looks like.

was to GM crops. However, although experimenting with human genetics is considered anathema in many societies, and is illegal in most, that prohibition is a cultural taboo specific to our particular time and circumstances. Again, we must avoid anthropocentrism by attributing the same reservations to alien societies.

Once a species embarks on enhancement technology, very rapid changes can be expected. We can glimpse the possibilities by reflecting on what may lie in store for humans, if the cultural taboos are eventually lifted.[3] Already many futurists are forecasting the onset of transhumanism, involving a combination of genetic improvement, prosthetics, life prolongation and neurological augmentation. Much of this is already happening. Average life expectancy has been increasing at an incredible three months per year for over a century, merely as a result of basic public health and medical advances. Prostheses will soon approach and even exceed the quality of the natural items; for example, artificial limbs, then eyes, will soon be wired directly into the brain. Implanted microchips will operate electronic systems in our environment. These devices will be augmented by organic body parts grown from stem cells, in some cases deliberately engineered for improvement. Hybrid 'natural-artificial' or 'organic-mechanistic' systems will be developed, opening up a much bigger space of possibilities than exists in the biological realm alone, and turning the fictional concept of the cyborg into a reality. It is reasonable to expect that any intelligent species that discovers biotechnology, nanotechnology and information technology will eventually employ them to boost its physical and mental capabilities. At that point, there could emerge a Utopia in which computer-designed beings enjoy the best of biological qualities without the inconvenience of illness or early death, flawed memory and poor reasoning. It is easy to imagine an alien society attaining this idyll after only a few centuries of science and technology.[4]

However, even after all the above-mentioned improvements, the enhanced beings would still be recognizably biological organisms – which brings me to what is probably the greatest fallacy of all in expecting Mekon-like extraterrestrials, or indeed any 'flesh-and-blood' aliens. When contemplating alien civilizations we have to consider much longer time frames than just the few centuries it may take for the above-mentioned technological advances, in which case an even more radical possibility must be confronted. 'Intelligence' on Earth is normally associated with hominids, and perhaps in a more limited way with cats, dogs, dolphins, whales, cephalopods and birds. It is clear, however, that intelligent decision-making and behaviour need not be the exclusive preserve of animals. Indeed, it need not be restricted to biology at all.

ARTIFICIAL INTELLIGENCE

In 1950, Alan Turing published a groundbreaking paper in the journal *Mind* with the provocative title 'Can machines think?'[5] Turing extrapolated from first-hand experience of the fledgling computer industry to envisage a time when a manmade electronic device could mimic human responses so convincingly that we would attribute consciousness to it. A few years later, Isaac Asimov developed this theme in his classic novel *I, Robot*. By the 1960s the subject of artificial intelligence, or AI, was appearing on the agendas of commercial and university research, and was also seeping into popular culture. In Stanley Kubrick's film *2001: A Space Odyssey*, the supercomputer HAL is portrayed as an intelligent being in competition with humans. By the time *Star Wars* was released, viewers had become accustomed to the idea of intelligent robots fighting and working alongside humans as equals, or even as superiors. Today, we have little difficulty accepting that computers can outperform humans in many mental tasks. It doesn't require too much stretch of the imagination to believe that, within a few decades, they will outsmart us in *every* way. Very soon intelligent machines, computers and robots will take over many functions now being performed by people. The same could be true of any intelligent alien species.

To guess how this might play out on an alien planet, we can consider some of the developments in artificial intelligence taking place on Earth. The adult human brain contains about a hundred billion neurons, networked so densely that the average neuron has over 1,000 synaptic connections, some much more. Typically a neuron will fire up to 500 times a second, so if the whole brain were firing flat out (a purely imaginary prospect I might say), there would be 40 trillion synaptic firings per cubic centimetre of grey matter – that's 40 teraflops in computer jargon. How do computers compare? Coincidentally, today's supercomputers could also achieve about 40 teraflops per cubic centimetre if every switch fired at once. The big difference is that the computer would consume several megawatts to do it, whereas the brain gets by on three meals a day. Taking the brain as a whole, it executes about 10,000 trillion operations per second (the number is a bit ill-defined). The fastest supercomputer achieves 360 trillion, so Mother Nature is still ahead. But not for long. If Moore's law holds up, the computer industry could be touting exaflops (that's a million trillion operations per second) by 2020 and zetaflops (a billion trillion flops) a decade later. Clearly, measured in terms of crude processing power, high-performance computing is set to soon overtake the human brain. Once that line has been crossed then, in principle, artificial intelligence could rival human intelligence. But there are huge caveats. For a start, the neural architecture of a brain is totally different from the wiring layout of a computer. Moreover, the software for managing all those frenetic flops in such a way as to mimic human-like intellect is not at all understood. And then there is the question of all that sensory input and motor control.

Rather than implementing AI by trying to build a cleverly programmed silicon brain from scratch, another approach suggests itself. Why not use all that amazing computing power to *simulate* a brain? The distinction here is crucial. Instead of using a computer to *mimic* a brain, the computer is programmed to model the actual goings-on inside a *real* brain, from the bottom up. In effect, the computer becomes a virtual brain (as opposed to an artificial rival to the brain). It is a tantalizing prospect.

Would it be possible in the near future to effectively model the *entire* human brain on a supercomputer? Yes, according to computational neuroscientist Henry Markram, who heads the so-called Blue Brain Project in Lausanne, Switzerland. In his ambitious scheme, each neuron is modelled mathematically by equations with up to 500 variables, leading to accurate predictions of the behaviour of single neurons undergoing electro-chemical stimuli. Real neural architecture is then adopted as a blueprint for virtually 'wiring together' the simulated neurons, thus creating a neural network *in silico*. If the job is done properly, the patterns flowing around the network in the computer simulation should accurately mirror the patterns flowing around a real brain. In a pilot study, 10,000 neurons were digitally linked and used to model a component of a mammalian cortex, with convincing results. This was the trigger that led Markram to scale up and tackle the entire mouse brain as the first step on the road to the human version. His goal is to capture a hundred trillion synaptic connections in the computer simulation! That is currently well beyond the computational resources of the project, but with the advances in computing expected over the coming decades, Markram's dream could be realized by the middle of the century, if not before.

The Blue Brain Project raises a fascinating philosophical question. One of the deepest scientific mysteries is the nature of consciousness: specifically, how does the brain create it? What does it take in the way of swirling electrical patterns to make a thought or a feeling or a sense of self-awareness? Nobody has the slightest idea. But if Markram's simulation is accurate enough then, *by definition*, his computational system will not only be intelligent; it will be a conscious, feeling, sentient *being*. In short, just what Turing had in mind. Of course, we may still be no nearer to solving how the brain actually does it, that is, we may not be able to discern precisely *what* features of neural circuitry are responsible for consciousness, although it is very likely that we would learn a lot from being able to simulate the phenomenon step by step. There is an obvious ethical issue here. If Markram's silicon super-brain is a conscious agent, it will surely deserve some rights. Fiddling with the programming to figure out what makes 'it' tick may rightly be considered immoral. I should stress that the Blue Brain Project is not some ghoulish attempt to fashion a virtual Frankenstein's creature. Rather, the prime motivation is to provide insights into precisely what goes wrong at the neuronal level in malfunctioning brains, such as those with Alzheimer's or Parkinson's disease.

When combined with genome analysis, real brain simulation will open up astounding possibilities for designing, modifying and creating thinking entities with powerfully amplified capabilities of reasoning, artistic appreciation, ethical standards, problem-solving ability – you name it. If stem cell research matches the advances in genomics and computing, it will one day be possible to grow in the proverbial vat not just spare kidneys and livers but entire brains, enhanced by genetic modification, and designed in advance by computational neuroscientists to meet certain performance criteria. The next step will be to merge these designer brains with non-biological materials and circuits, thus augmenting what can be achieved by biology alone. As in the case of nanotechnology and biotechnology, a fusion of biological and non-biological neuroscience will soon eliminate the distinction between what is a brain and what is a computer. These systems may be deliberately created to omit certain human-like qualities – for example, moodiness or impatience or jealousy – but they will attain such a high level of expertise and competence that we will come to trust their judgement on an ever wider range of decisions.

It is inevitable that at some stage these designed and fabricated agents would have to be given a measure of autonomy to function at maximum efficiency, because we mere humans would not be able to keep up with them intellectually. In science fiction, this step is often portrayed as the machines 'taking over' from humans, with the implied threat that the machines may then turn on us and even annihilate us. But this is to fall into the trap of anthropomorphizing machine intelligence. There is no particular reason why human and computer agendas could not be harmonized. Free of primitive Darwinian urges such as fight-and-flight, disgust and the need for procreation, autonomous computers are unlikely to see humans as threatening or in competition with them (unless, of course, we try to switch them off).[6]

What might the computers'/robots' agenda be? Because we are now into extremely speculative territory, this question is almost impossible to answer. Initially, humans would create these machines to assist their own endeavours, and this the machines may continue to do, but in due course they would find better things to occupy their time, about which we can only guess. Assuming the machines at least wish to secure their own survival (as individuals, not through procreation) and extend their reach in some way, they will need their own tools. Like humans before them, the computers will make machines to carry out a variety of tasks. Some of these machines might be similar to ours – motors to move hardware about, dynamos to make electricity, telescopes to survey the heavens and search for threats such as incoming asteroids. Others, however, would be biological. Microbes to sequester and process minerals needed for construction is an obvious example. Other microbes might be designed to change the physical conditions of the machines' environment. The machines might also design and manufacture mesoscopic (small, but not microscopic) or even macroscopic organisms to fulfil specialist functions such as maintenance, exploration and observation. If the machines/computers were sedentary, these complex organisms could be their roving eyes and ears, roaming the planet or being dispatched to other planets to gather information.

For hundreds of thousands of years humans manipulated their world using simple tools to improve their chances of survival. At first progress was very slow, and the tools were limited to clubs and spears. With the development of language, settled communities and agriculture, the pace accelerated, leading to the bow and arrow, the use of metals, the plough and the wheel. Before long, the Industrial Revolution happened, followed by the atomic age, the space age and the computer age. Throughout this great span of history, humans used technology to improve their well-being. But we can now foresee a tipping point when this longstanding relationship between the biological and non-biological realms will become inverted. Instead of life forms such as humans designing and making specialized machines, machines will design and make specialized life forms. The baton of intelligence – the all-important 'I' in SETI – will have been well and truly passed to the machine realm. Intelligent biological organisms would henceforth exist in a purely subordinate role. Because of the greater robustness of machine intelligence, its survival prospects are far superior to that of humans, or of any other flesh-and-blood entity. Machines can easily be made immortal, by replacing their parts with spares when they wear out. They can also be merged to make bigger and better machines, and can function under a wide range of physical conditions. All in all, machines offer a far safer and more durable repository for intelligence than brains.

My conclusion is a startling one. I think it very likely – in fact inevitable – that biological intelligence is only a transitory phenomenon, a fleeting phase in the evolution of intelligence in the universe. If we ever encounter extraterrestrial intelligence, I believe it is overwhelmingly likely to be post-biological in nature, a conclusion that has obvious and far-reaching ramifications for SETI.

I'VE SEEN ET, AND IT'S AN ATS

Human intelligence is no more than a few hundred thousand years old, depending somewhat on definition. In a million years, if humanity isn't wiped out before that, biological intelligence will be viewed as merely the midwife of 'real' intelligence – the powerful, scalable, adaptable, immortal sort that is characteristic of the machine realm. Thereafter, machine intelligence will accelerate in power and capability until it hits fundamental bounds imposed by the physical environment, whatever they might be. And at that stage, the self-created godlike mega-brains will seek to spread across the universe. By the same token, we can expect any advanced extraterrestrial biological intelligence to long ago have transitioned to machine form. Should we ever make contact with ET, we would not be communicating with Mekon-like humanoids, but with a vastly superior purpose-designed information-processing system.[7]

I have, unfortunately, lapsed into sloppy terminology over the last few pages. As I described earlier in this chapter, the distinction between living and non-living, organism and machine, natural and artificial, is set to evaporate soon. To call the alien entities 'computers' and 'machines' is misleading. They might, for example, be hybrids with organic and inorganic components intermingled, so they would not be living organisms in the usual sense of the word, but they would not be inanimate either, because they could grow and regenerate components biologically. It's hard to know what to call such entities, because they are beyond human experience. Their characteristic property is that they are the product of design, originally (in the case of future Earth) by humans or (in the case of an alien civilization) their extraterrestrial counterparts. Later they would be self-designed and redesigned. They would be systems that grow, improve and adapt, not by some long-winded Darwinian mechanism, but through their own intellectual creativity. The best term I can come up with is the horrendous-sounding 'auto-teleological super-systems' (ATS); the adjective implies the property of goal-oriented self-design. Because manipulation by design is so much more efficient than Darwinism, the self-design process, once triggered, is therefore likely to be very fast, greatly increasing the likelihood that the 'I' in SETI is dominated by ATSs.

As I write these outlandish speculations, I find myself curiously depressed, nostalgic-in-advance for the personal identity that is so much a characteristic of human experience. Each of us has a unique sense of self, a feeling of being part of, but separate from, a community of other sentient beings, and the wider universe. How the human brain generates the impression of separate self-identity, and the subjective experiences that go with it, is still a complete mystery, as is the evolutionary pathway that led to it. However, there is no good reason for an ATS to possess a personal identity in anything like the same way.[8] The power of computers is that they can be linked together, without much protest, to share tasks and pool resources. Unlike brains, which are discrete entities, computers can be networked, merged, reconfigured and expanded seemingly indefinitely. Think of a search engine like Google, which has a global reach via the internet and distributes its operations to computer clusters located in many places around the world. A powerful computer network with no sense of self would have an enormous advantage over human intelligence because it could redesign 'itself', fearlessly make changes, merge with other systems and grow. 'Feeling personal' about it would be a distinct impediment to progress.

It isn't hard to envisage the entire surface of a planet being covered with a single, integrated information-processing system. In fact, some futurologists picture the entire surface of a Dyson sphere being devoted to a gigantic pulsing mega-brain (like Plate 14 perhaps). Robert Bradbury has coined the term 'Matrioshka brains' for these awesome entities.[9] Even if someone could work out how to link and merge human brains and experiences into a sort of World Wide Web of Wisdom, most of us (at least in Western culture) would be appalled by the prospect of losing our selves in a vast amorphous mental space. The considerable literature on 'uploading', a fantasy in which the contents of ageing brains, and by implication their associated conscious selves, are transferred to a computer, later to be downloaded into new brains, is appealing precisely because of the implied continuity of the self and the promise of immortality.

If biological intelligence is destined to 'hand over' to ATS intelligence, where will it all end? Well, even these mind-boggling mega-brains are still subject to the laws of physics, such as the finite speed of light. A computer that enveloped the Earth, or a Matrioshka brain, might have some wonderful thoughts, but its train of thought would necessarily be shackled by the significant fraction of a second it takes for information to be shunted from one region of the system to another. In effect, a monster ATS would be dazzlingly brilliant but relatively slow-witted. The same limitation is even more severe for a larger-scale system, such as a galactic Google, where delay times of 100,000 years would impose a stringent limit on data recovery and hence on the speed of thought.

So is that it? A universe dominated[10] by enormous but plodding intellects? Perhaps that is as far as machine intelligence can go. But if certain recent developments in information-processing are right, there might be a way to go further, a way that would create a type of intelligence that is alien even by the standards of an ATS.

QUANTUM COMPUTERS AND QUANTUM MINDS

The basis of all digital computation is the binary switch, a device that can be either on or off. It needn't be a mechanical switch: normally it is an electronic component that has two states. If off stands for 0 and on for 1, a network of switches can process digital information simply by flipping en masse to convert input sequences of 0s and 1s to output sequences. The details are unimportant for the purpose of this discussion. The speed of computers is limited by the rate that the switches can flip and by how fast the electrical (or optical) signals encoding the 0s and 1s can pass from switch to switch. Ultimately the speed of light imposes an absolute limit, but by making the system smaller it can run faster. The light travel time across a typical personal computer microchip is less than a picosecond (a trillionth of a second); if the chip were more compact, the processing speed could be higher. But shrinking the chip brings its own problems. One of these is heat. Every time a switch flips, even a non-mechanical one, heat is generated, and this has to be dissipated somehow or the chip will melt. Physicists know that the heat produced by today's microchips can in theory be very substantially reduced, so in the long term heat may not be the dominant issue. But a tougher problem awaits, one that is not so easily evaded. As the basic switch size approaches atomic dimensions, the physical properties of the circuits are more and more subject to the perturbing effects of quantum fluctuations.

Quantum mechanics is the theory that describes the weird behaviour of atoms and subatomic particles; I touched on it in Chapter 7. It differs radically from Newton's mechanics, which apply to everyday-sized objects like billiard balls and bullets. The key characteristic of quantum systems is uncertainty. Let me give a simple example. If a gun is fired at a target, the bullet follows a well-defined trajectory through space. Repeat the experiment, under identical conditions, and the second bullet will follow the same trajectory as the first. In such cases, nature is deterministic; knowing the initial conditions plus the laws of mechanics enables one to correctly compute the trajectory in advance. Simply put, the system is predictable. Quantum mechanics is a very different kettle of fish, however. An electron or atom fired at a target may follow many different trajectories and hit the target at many points. If the experiment is repeated, *even under identical conditions*, it will not normally produce the same outcome.

Not all everyday phenomena are predictable. Tossing a fair coin produces heads or tails with 50 per cent probability, but it's impossible to know the outcome of an individual toss because the result is so sensitive to unknown forces acting on the coin. Quantum uncertainty is quite unlike that. It arises not because we are ignorant of all the forces determining the outcome, but because the system is *intrinsically* indeterministic. Expressed more graphically, even nature doesn't know what will happen case by case. From the point of view of computation, unpredictability is a disaster. What good is it if 1 + 1 = 2 on the first attempt and 3 on the second? If the components in a computer chip are shrunk towards atomic size, quantum uncertainty lies in wait to compromise the performance.

While these weird quantum effects seem to scupper all hope of reliably computing at the atomic level, it turns out the converse might be true. When a tossed coin has fallen, even if we don't look at the outcome there is no doubt that the upturned face is *either* heads *or* tails. By contrast, quantum mechanics permits an atom to be in the equivalent of *both* heads *and* tails at once, a ghostly hybrid state that is projected into concrete reality only after an observation is made!

Furthermore, this admixture can vary continuously from all heads, through mostly heads plus a bit of tails, to more of tails than heads, and so on, as far as all tails.[11] Translated into the context of a computer chip, quantum mechanics says a given switch isn't generally *either* on *or* off, but a bit of both. The closer the switch gets to atomic dimensions, the more this 'superposition' property is manifested. And therein lies the secret of the much-sought-after quantum computer, a device I mention in Chapter 5 as a test of alien technology. Physicists believe they can turn a sin into a virtue by harnessing superpositions to carry out computations, and if it is done right, the results can be completely free of uncertainty.[12]

The idea of a quantum computer has captivated the imagination of scientists and the computing industry alike, and is now the subject of a major international research effort.[13] The reason for the surge in activity is the discovery that a quantum computer could solve certain problems not merely a lot faster than a conventional computer, but *exponentially* faster, representing an advance over current supercomputers as great as that of the electronic computer over the abacus. A quantum computer that fully controls a mere 300 atoms could in principle store more bits of information than there are particles in the entire observable universe. That doesn't mean we could build a computer as powerful as the universe with only 300 atoms, though. Storage is one thing, processing is another. Quantum states are incredibly fragile, and any extraneous disturbance degrades their performance. The secret to successful quantum computation is to allow the system to evolve with time while isolating it as much as possible from its surroundings, and to compensate for accumulating disturbances with error correction techniques and redundancy. All this is a matter of engineering, and a variety of tricks is currently being investigated, such as trapping individual atoms in magnetic fields at ultra-low temperatures. What nobody knows at this stage is whether error correction can ever be made perfect, or whether there are deep principles of physics that impose a diminishing-returns penalty, implying a fundamental limit on the power of quantum computation. The experts say that doesn't seem to be the case, but so far they have managed to harness only a dozen or so atoms in concert. An advanced alien technology might be able to manufacture a near-perfect quantum computer that would be physically very compact (say, the size of a car) yet have staggering information-processing power, perhaps creating in a single lab a super-intelligent machine possessing the same capability as a conventional computer that covers an entire planet.

If quantum computers are as feasible as their proponents claim, then we might very well expect ET to *be* a quantum computer. If so, where might it be located? It seems unlikely that an EQC (extraterrestrial quantum computer) would reside on a planet. Random disturbances – the enemy of quantum computation – derive from heat, so locating the EQC in the coldest possible environment available makes sense. Interstellar or intergalactic space would be ideal. In any case, planets are dangerous places in the longer term, because of comet impacts, supernova explosions, instability of the host star, orbital irregularities and so forth. A dark quiescent void would be much better, so long as an energy supply and some raw material are available. An asteroid propelled into intergalactic space may suffice for the latter; the former might be met by cosmic rays.

Mulling over these fantastic ideas about the outer reaches of intelligence, I keep coming back to the same thorny issue. Why would such an entity bother to contact us? What could we possibly say to it? In fact, it is not at all clear to me that an intelligent quantum computer would have much interest in the physical universe at all. So what would an EQC do for thrills? By definition, this entity resides not only in physical space but in cyberspace. Even supposing it possesses emotions, it would be much more likely to experience gratification in its own world of virtual reality, exploring an inner intellectual landscape that could be incomparably richer than the physical landscape (or spacescape) that surrounds it. But by retreating into cyberspace, the EQC would effectively disconnect from the universe that humans inhabit, apart from the minimal requirement of maintaining its own existence (such as paying the electricity bills and replacing faulty parts). Once it had secured safety, stability and an extreme degree of isolation, its own future would be guaranteed for trillions of years, barring unforeseen accidents that couldn't be dealt with by automatic repair mechanisms. Quite what it would choose to do with itself is utterly beyond us, although some commentators have suggested that super-advanced intellects of this sort would spend most of their time proving ever more subtle mathematical theorems. I confess this seems to me a rather narrow vision of thrill-seeking, but it may be that an EQC would rapidly exhaust all other possible experiences. It is known that mathematics possesses unlimited diversity and infinitely many surprises, so no matter how long the EQC extends its intellectual adventure, there will always be one more mathematical relationship for it to prove and admire.

Retreat into cyberspace is probably the most dispiriting resolution of the Fermi paradox. I hope it is wrong, for it would mean not only that biological intelligence is a transitory phase, but also that engagement with the real physical universe is transitory. From the point of view of SETI, however, what matters is whether an EQC produces an observable footprint in the real, physical universe. According to the basic physics of quantum computation, the core information-processing uses essentially no energy. But to maintain the delicately controlled conditions for that processing to work would entail elaborate equipment and a power source. If, as I suggested, the power requirements could be met from cosmic rays in intergalactic space, it is hard to imagine an EQC would ever be detectable from Earth. But if for some reason the peripheral equipment for quantum computation demands very much greater power, then there may even be quantum Matrioshka brains out there somewhere, enveloping stars or rotating black holes. Although we would never expect to receive messages from these quantum cyber-minds, their presence might nevertheless have a noticeable impact on the physical universe that supports them.

The new SETI programme I have outlined shifts the emphasis away from seeking messages for mankind using radio telescopes to the less ambitious goal of simply trying to identify signatures of intelligence through the impact that alien technology makes on the astronomical environment. To guess what to look for, I have used our best understanding of modern science and extrapolated into the future. But that strategy is open to the recurring charge of anthropocentrism. It is entirely possible that alien technology would involve things we haven't even dreamed of, and would produce physical effects that have yet to make it to anyone's list of things to watch out for. In pursuing new SETI, it is important to remember the adage: expect the unexpected.

New SETI is not intended to replace traditional SETI but to complement it. Even if my wild speculations about quantum Matrioshka brains and other exotica are correct, not all extraterrestrial intelligence will have attained, or ever will attain, such an advanced state. There is more likely to be a spectrum of intelligence, from alien communities not yet entering the age of technology, through biological organisms with the capability of signalling using radio, computer-dominated societies that retain (and maintain) biological communities, to full-blown cyber-intellects. It would be unwarranted to suppose that *none* of these hypothetical communities, at any level of advancement, will ever transmit messages, or build beacons or monuments intended to make a statement to their cosmic cousins. And while there is even a remote chance that someone, somewhere, wants to attract our attention, we should go on looking, for the consequences of success would be truly momentous.

9
First Contact

The societal and cultural impact might be more akin to the consequences of a religious revelation.

Stephen Baxter[1]

THE POST-DETECTION TASKGROUP

In 2004, Ray Norris, a radio astronomer in Sydney, Australia, asked if I'd consider taking over from him as Chair of the SETI Post-Detection Taskgroup. This curious body was constituted by the SETI Permanent Study Group of the International Academy of Astronautics (IAA), a scientific institution devoted to fostering the development of astronautics for peaceful purposes, with participation by over sixty countries. The brief of the Taskgroup, in a nutshell, is to prepare for the Big Day. Even if the chance of humankind being contacted any time soon by an extraterrestrial civilization is remote, it makes sense to think through some of the implications should it happen. We don't want to be caught on the hop. I agreed to serve, and after being duly elected I convened a meeting in February 2008 at the Beyond Center, Arizona State University.

The Taskgroup is only a think tank; it has no legal status and no teeth to impose its policy recommendations on anybody. Its members are nominated, and elected to the Permanent Study Group. They include leading SETI scientists and activists, representatives of the media, two lawyers, a philosopher, a theologian and two science fiction writers. The Deputy Chair, Carol Oliver, bridges the two cultures by having a background in print journalism as well as many years' experience as a SETI researcher in Australia. The primary purpose of the Taskgroup is to provide a resource for astronomers generally, and SETI researchers in particular, about post-detection issues. The Taskgroup's protocol was constructed in 1996 by the astronomer John Billingham, and is available on the web.[2]

In the event that a putative signal is detected, it would be the Taskgroup's job to counsel the parties concerned. If the protocol works as advertised, then the first task would be to urge the discoverer to subject the data to careful checking and evaluation. If the signal eventually proves genuine, then our advice would be for full details to be disclosed to the astronomical community first, especially the International Astronomical Union (IAU), the premier institution in the field of astronomy, which enjoys good links with many other scientific and government organizations around the world. The IAU would then be able to disseminate the news to the United Nations and other key bodies. In the early days this was to be by telegram – a quaint touch. Today it would be done electronically. The discoverer would also be advised to inform the government of the country in which the radio telescope is situated. Following that, she or he would be free to call a press conference or make a public announcement in some other way, should they so choose. In practice of course, it might be messier than this. The discoverer may be deliberately uncooperative or overawed and disoriented by the magnitude of the events. There may be more than one person and one country involved. The news might leak out ahead of the formal diplomatic steps (I shall have more to say on that below). Also, there is nothing to stop an astronomer who detects a signal out of the blue from going straight to the press or to her or his government, or any other organization, bypassing our Taskgroup altogether. However, the most likely scenario is that a detection event comes from within the SETI community, and in that case the Taskgroup's protocol is likely to be adhered to, and its advice heeded. Anyway, that's the theory.

As a result of my elevated status in the SETI world, I began to think more carefully about post-detection. What would happen if, suddenly, we found we were not alone in the universe? How would the discovery play out? After all, this would be a scientific finding without parallel, with ramifications going far beyond astronomy. I like to enliven my after-dinner speeches with the quip that if ET calls on my watch, I will be among the first to know for sure that there are aliens out there. I would be standing at a pivotal point in history, able to play an active part in the outcome. It gives me and my fellow Taskgroupers an awesome responsibility.

Reflecting on the subject of first contact, I realized that my preconceptions had been shaped largely by science fiction, in which the aliens are usually the bad guys. From *The War of the Worlds* through *Quatermass* to *Independence Day*, extraterrestrials are portrayed as a sinister threat to humanity. Only a handful of stories, like *Close Encounters of the Third Kind* and *Contact*, buck the trend. Even when the aliens don't show up in the flesh, contact stories rarely end happily for humanity. In Fred Hoyle's *A for Andromeda*, for example, a radio message received from a very distant star system contains information needed to reconstruct an alien, with potentially dire consequences. Hoyle's thesis, presented in 1961 in the form of a British TV drama, holds a chilling warning for the Taskgroup: can we trust ET not to dupe us? An alien civilization might not be explicitly hostile to humans. It could regard us as mildly useful, but ultimately 'in the way' and of little relevance to their grand scheme. They might enlist our help, then elbow us aside. Hoyle's brilliant plot, written just after the inception of Project Ozma, demonstrated that it isn't necessary for aliens to travel across space physically in order to colonize another world. All they need is to beam the required biological information to trusting scientists, and persuade them to incubate copies of the extraterrestrials in a sort of long-range version of *Jurassic Park*. To work well, the fabricated beings would require some adaptations to the local biology, which in the case of *A for Andromeda* took the form of the actress Julie Christie.

So much for the fears. What about the hopes? SETI researchers are buoyed by the expectation that contact with an advanced alien civilization has the potential to bring untold benefits to mankind. Being in touch with ET would expose our civilization to accumulated cosmic wisdom, and open the way to technological marvels, deep scientific insights and entry to the Galactic Club. Those who take a rosy view of aliens dismiss the scary Hollywood image as overly anthropocentric, and point out that any beings that have overcome their own problems and survived for eons are unlikely to be innately aggressive. An alien civilization that goes to the trouble and expense of actively trying to contact us would probably be highly altruistic. They would presumably be aware of the danger posed when a technologically advanced culture comes into contact with a less advanced one, and manage the interchange with sensitivity. Well, maybe. It is the Taskgroup's responsibility to weigh up all the pros and cons about first contact, and to formulate a plan of action, so there is some measure of consensus on what to do.

MEDIA FRENZY

Let me focus on the first step following the detection of a putative signal – checking the authenticity. In the case of traditional radio SETI there is a tried-and-tested protocol for a real time 'detection event' (as opposed to something uncovered later in recorded data), which is designed to eliminate false alarms such as equipment malfunction and manmade signals. As I explained in Chapter 1, a key check is to obtain verification from an independent radio observatory. That takes time, and things don't always run smoothly. On one occasion in 1997, a strong narrow-band signal from space was detected at Green Bank, West Virginia, during a SETI run. A check of all known satellites did not find a match, and by bad luck the back-up telescope at Woodbury, Georgia, was down. There was considerable excitement at Green Bank for a day or so before the signal was eventually identified as coming from a research satellite called SOHO. The interpretation was complicated by the fact that the radio telescope was not actually pointing at SOHO (which is orbiting near the sun). By a quirk of radio physics, its signal had been picked up in weakened form edge on, in the so-called 'side lobe' of the dish.[3]

The fact that it may take days to be sure that a signal is not manmade raises a very serious problem for managing the post-detection agenda. A message from an alien source would be an event of unprecedented significance. Any hint of a positive result from a SETI project could immediately trigger media frenzy, and events might soon spiral out of control. All it takes is one intemperate remark by an observatory janitor, and the story will spread like wildfire. Even if nobody actively spills the beans, a tight-lipped silence in the face of a routine press enquiry might well be interpreted as some sort of cover-up. In the case of the SOHO satellite detection, the press got hold of the story even before the identification was made.[4] Fortunately the reporter concerned acted responsibly and waited for more data before rushing into print. But not all members of the media can be relied upon to be so restrained, given the chance of the scoop of a lifetime.

The Taskgroup has deliberated in depth over how to manage the situation following a putative signal, especially in the light of the revolutionary changes in communications and media that are occurring, from the use of the Web and Web 2.0 technology, mobile phones, Twitter, Facebook, etc., all of which are transforming the speed and manner in which information, discoveries and opinion are disseminated. Two members of the Taskgroup, Seth Shostak and Carol Oliver, have drawn up an Immediate Reaction Plan to minimize the amount of misinformation promulgated in the wake of claimed ETI detections.[5] They note that because SETI is carried out openly and with no policy of secrecy, word can leak out very fast. The media will in all likelihood run with the story even before the initial scientific checks have been completed. 'The story will break before it's a story' is the way they put it.[6] As a result of their report, the Taskgroup has set up a password-protected website so that members can confer and post information, at a time when publicly accessible SETI websites are likely to be paralysed by hits.

The fundamental problem concerning media management derives from a deep cultural rift between the world of science and the world of news and commentary. Because SETI astronomers are professional scientists, rigorous checking is an essential part of their training, and they want to be sure of their ground before making a definitive statement. History has shown that when scientists run to the press with sensational claims that haven't been properly checked, the outcome is very damaging to the credibility of science itself, not to mention the reputations of the scientists involved. A salutary lesson in how *not* to handle the media comes from the now largely discredited claim for cold nuclear fusion. This story broke in 1989 when two physicists said they could produce nuclear fusion reactions in what was basically a test tube on a bench top, by doping the metal palladium with deuterium. Had they been right, all the world's energy problems would have been solved at a stroke. They held a hasty press conference, and the media understandably had a field day. Cold fusion became the big science story of the year.[7] It took many months for laboratories around the world to test the claim, and find it wanting. The two scientists themselves were hounded by the press and went into hiding. Today, a handful of labs continue to work on cold fusion out of curiosity, but very few physicists believe it will ever amount to much. The lesson from that debacle is that it is wise to exercise restraint when dealing with the media about discoveries that carry sweeping implications for society.

In the case of SETI, the problem is far more acute. The scientists might be sitting on the biggest story in history. Once word got out, mayhem could ensue. The astronomers might show up for work only to find their observatory besieged by journalists, film crews and members of the public, some of them excited and others frightened. There would have to be a police blockade, and protection for both the scientific and the technical staff – hardly an environment conducive to dispassionate analysis. Even normal modes of communication are likely to be disrupted as lines become jammed by callers eager to check the rumours, computer servers become overloaded and hackers try to break into the system to get a sneak preview of ET's message.

It is in the nature of this type of investigation that false alarms greatly outnumber the real thing, so the above scenario might be played out many times, with the hullabaloo eventually subsiding as the story evaporates. A close analogy is the all too frequent announcement that civilization is menaced by an oncoming asteroid or comet. Thousands of small objects are on Earth-crossing orbits, and from time to time one of them scores a hit; the scars of their impacts can be seen scattered across the planet from Meteor Crater in Arizona to Wolfe Creek in Australia. The damage from an impact depends on the size and speed of the colliding object. A relatively rare impact of the power that wiped out the dinosaurs would probably annihilate humanity too, but these happen on average only once every 30 million years or more. Smaller events are more likely, but still have great destructive potential. For example, an asteroid one kilometre wide hitting Earth at 30 kilometres

(20 miles) per second might kill a billion people, from both the collision itself and the unpleasant aftermath (which includes wildfires, acid rain, sun-obliterating dust and a host of other nasty effects). There is roughly a one-in-a-million chance that such an event will happen next year.

For the past couple of decades, astronomers have been painstakingly cataloguing the orbits of the more dangerous asteroids, so that we at least have some warning of the next big impact. When a new asteroid or comet seems to be moving on an Earth-crossing trajectory, it is carefully monitored so its orbit can be determined precisely. As with SETI, careful checking takes time. In the early days following the discovery, the projected orbits are uncertain because of normal measurement errors. After the object has been followed for several days or weeks, the errors shrink enough that the astronomers can then work out whether it will or won't hit Earth. The most sensible strategy is to wait until the orbit has been properly determined, and only then, if there is still a clear and present danger, 'wake the President'.[8] But usually it doesn't happen like that. More often than not, the press get wind that a new object has been found that *might* strike our planet on the next orbital pass. It makes a wonderful scare story: 'Killer asteroid may wipe out life as we know it!' Headlines like that attract a lot of readers, particularly when Armageddon comes with a specified date. But there is a world of difference between predicting that an object *will* hit, and being unable to rule out that it *won't*. The known uncertainty in the measurements lets astronomers work out the probability of a collision – typically it is about one in 10,000 when the object is first identified. Those odds can still seem frightening for such a major calamity, but another way of looking at it is that there will be thousands of apocalyptic scare stories appearing in the press before the one case when a collision *will* result.

THE BLANKET OF SILENCE FALLACY

Unfortunately, waiting to be sure has its own drawbacks. If scientists respond to a query about an asteroid impact or a SETI rumour with a simple 'no comment', the press and the public are all too ready to suspect a conspiracy of silence. People justifiably believe in a right to know, and are suspicious when scientists seem to be hushing up their findings, even if the motive is normal scientific prudence rather than a deliberate news blackout. Most members of the public just don't buy the 'trust us, we're scientists' line. Conversely scientists, concerned for their reputations and funding, can be fiercely critical of the media, whom they see as all too prone to scaremongering. The BBC science correspondent David Whitehouse was accused of crying wolf when, in 2002, he ran a premature news story about a possible cosmic impact on 1 February 2019. In response, Whitehouse hit back on the subject of scientists keeping mum: 'Who gives them the right to make such a decision? Who actually would make the decision? What would be their qualifications, their accountability? . . . The ethics of such a stance are unsupportable. There are other areas of science where the "they don't need to know" argument has been debated and discounted as unethical.'[9]

I personally believe the public does have a right to know, even if the news is bad – as soon as the situation is properly understood. I have yet to meet a SETI scientist who doesn't agree with this basic principle. There is no 'code of secrecy' in SETI, and certainly not among the Post-Detection Taskgroup's members; only a shared recognition of the need for caution when assessing any putative signal. The IAA itself is explicit (if a little turgid) about disclosure in items 3, 4 and 5 of the SETI Permanent Study Group's 1997 'Declaration of Principles Concerning Activities Following the Detection of Extraterrestrial Intelligence':[10]

3. After concluding that the discovery appears to be credible evidence of extraterrestrial intelligence, and after informing other parties to this declaration, the discoverer should inform observers throughout the world through the Central Bureau for Astronomical Telegrams of the International Astronomical Union, and should inform the Secretary General of the United Nations in accordance with Article XI of the Treaty on Principles Governing the Activities of States in the Exploration and Use of Outer Space, Including the Moon and Other Bodies. Because of their demonstrated interest in and expertise concerning the question of the existence of extraterrestrial intelligence, the discoverer should simultaneously inform the following international institutions of the discovery and should provide them with all pertinent data and recorded information concerning the evidence: the International

Telecommunication Union, the Committee on Space Research of the International Council of Scientific Unions, the International Astronautical Federation, the International Academy of Astronautics, the International Institute of Space Law, Commission 51 of the International Astronomical Union and Commission J of the International Radio Science Union.

4. A confirmed detection of extraterrestrial intelligence should be disseminated promptly, openly, and widely through scientific channels and public media, observing the procedures in this declaration. The discoverer should have the privilege of making the first public announcement.

5. All data necessary for confirmation of detection should be made available to the international scientific community through publications, meetings, conferences, and other appropriate means.

Even if the scientists are prepared to be open about their findings, can we trust governments to act in the same way? In a typical science fiction story featuring alien contact, government security services instantly spring into action, take control of the project, and impose a cloak of secrecy. The clampdown is justified for reasons of excessive paternalism ('People aren't ready for this yet'), or to gain advantage ('We might learn something amazing that will enhance our power'), or to prepare a defence ('We must build more nukes'). Well, if there *are* government plans to seize control of SETI following a positive result, they haven't yet come to the attention of the SETI community, in spite of several high-profile hoaxes and false alarms.[11] In fact, far from taking an unhealthy interest in the subject, governments worldwide seem to be completely indifferent. A member of the British House of Lords once asked me about SETI, but purely out of personal curiosity. In the US, Congress cancelled public funding for SETI in 1993, on the basis that it was a waste of money. That is hardly the action of a government that has a serious interest in 'contact'. As for secret government post-detection contingency plans, I have no doubt they are non-existent. When it comes to post-detection policymaking, the Taskgroup is it. In fact, we would actually *welcome* some input from politicians, or at least from a few elder statesmen.

'IT'S OFFICIAL – WE ARE NOT ALONE!'

Suppose the authenticity-checking process is complete, and the discovery holds up at, say, 99 per cent confidence level (scientists never claim 100 per cent certainty about any discovery). The next step is for some sort of official announcement to be made. How should that be done? The manner will depend critically on the precise nature of the discovery. In my mind, there is a world of difference between the Holy Grail of SETI – picking up a directed message from an alien civilization – and the less dramatic but far more likely case of our simply obtaining incontrovertible evidence for some sort of alien technology. The latter case would be far easier to handle. If an astronomer were to spot something weird, which on closer inspection bore all the hallmarks of artificiality, then I believe it should be announced just like any other major astronomical discovery. During my career, astronomers have found a range of extraordinary new objects – quasars, pulsars, black holes and gamma ray bursters, to name just a few. Finding an 'intelligently modified object' in space would extend this list of mind-expanding findings. It could be a beacon (see Chapter 5), a sign of astro-engineering (Chapter 6), or simply a radio or light source lacking a plausible natural interpretation. All one could conclude with confidence from such an observation is that some form of intelligence had been at work elsewhere in the universe. Ideally a press conference would be arranged to coincide with the publication of a peer-reviewed paper in a reputable scientific journal, a process that typically takes some months.

There is no doubt that an announcement of an intelligently modified object in space would cause a sensation. When President Clinton stood on the White House lawn and said that NASA scientists had evidence for life in a Mars meteorite (see p. 61), the world's journalists were electrified by the news. Presenting evidence for *intelligent* life would be an order of magnitude more startling. For a few weeks, the story would run and run. Scientists would be pursued for interviews, commentators would offer impromptu assessments, and the blogosphere would buzz with half-baked theories. But after a while the newsworthiness would begin to fade, and the media would return to their usual fare of politics, sports and celebrity trivia. Life would carry on as before. The vast majority of people would go about their daily affairs with only a residual interest. It would, after all, make no difference to the price of beer or the outcome of the next big game: it would merely be a scientific curiosity.

Over the longer term, however, the discovery would have disruptive effects at many levels. History teaches us a lesson here. When Copernicus deduced that Earth goes around the sun it was considered a dangerously revolutionary discovery, in both the literal and metaphorical sense of the term. At that time, the controlling power *was* interested in suppressing scientific truth. That power was not a national government, but the Roman Catholic Church, which regulated almost every facet of European society, including information and education. What the Church feared was not riots or panic in the streets as a result of Copernicus' cosmic revelation; rather they foresaw the weakening effect it would have on their version of Christianity. Of course, they failed, and the heliocentric model of the solar system soon became accepted. And life continued normally; peasants still collected the harvest, noblemen still hunted and made war, and scholars (including within the Church) quietly assimilated the new cosmology. Four centuries later, what can we say about Copernicus' theory? There is no doubt that it fundamentally changed the way human beings think about themselves and their place in the universe. Each succeeding generation built on it and expanded humanity's view of the cosmos to encompass not merely our solar system, but a volume a thousand trillion trillion trillion times greater. Even today, for most practical purposes Earth might as well be at the centre of the universe. But the knowledge that our planet is a fragile, pale-blue dot in the vastness of space permeates our world view and exerts a subtle influence on our lives in a thousand different ways.[12]

A similar reception greeted the publication of Darwin's theory of evolution. The claim that humans had 'descended from apes' (a popular but very inaccurate description of the theory) caused shock and outrage in some quarters. It was certainly a 'big story' by Victorian standards. The Church was no longer powerful enough to suppress the truth, but it did put up a spirited resistance in some quarters before conceding defeat. Yet, once again, the vast majority of people went about their daily lives as before, assimilating the ideas at their own pace. There was no civil unrest, no public outpouring of despair, and no euphoria. One hundred and fifty years later, however, few would deny the powerful significance of Darwin's theory. Knowing that humans are a product of billions of years of natural selection – that you and I are an integral part of nature and not the product of special creation – colours our attitudes to our fellow human beings and animals. Today, when we address the question 'What does it mean to be human?' and reflect on our place in nature, our biological pedigree forms an indispensable backdrop to our thinking.

If we ever do discover unmistakable signs of alien intelligence, the knowledge that we are not alone in the universe will eventually seep into every facet of human enquiry. It will irreversibly alter how we feel about ourselves and our location on planet Earth. The discovery would rank alongside those of Copernicus and Darwin as one of the great transformative events in human history. But it would take many decades for people to adjust and for the full import to sink in, just as it did for heliocentric cosmology and biological evolution.

INTERCEPTING INTERSTELLAR E-MAIL

When Frank Drake embarked on Project Ozma, his aspiration was not merely to answer the question 'Are we alone?' but to establish actual contact with extraterrestrials. In spite of the big error bars in his eponymous equation, Frank remains upbeat. It's tempting to suppose that if an alien interstellar radio transmitter is on the air, Frank and his team of astronomers will find it within a few decades. If he is right (and you have to be an optimist in this subject), then we might soon be confronted by an alien message *with content*. For reasons I explained in Chapter 5, the radio signals are unlikely to be directed at earthlings specifically. Rather, they would be something coming our way by chance; we would in effect be eavesdropping on someone else's conversation, or intercepting their e-mail. Although it's hard to see how we could possibly decode the content, a great deal could be learned just from the structure of the signal. For example, we could locate the transmitter. If it turned out to be relatively close, we would have antennas powerful enough to send a decent-strength signal to 'them'. We could also look for the intended recipient civilization (presumably in a part of the sky antipodal to the transmitter), and target that region too in our search for signals.

It might even be possible to determine the informational richness of the message without decoding the actual content. This is because data-rich messages satisfy certain statistical criteria irrespective of the meaning being conveyed. A simple example illustrates this. If I send a message and then repeat it, the redundancy reduces the total content by a factor of two (because half the data bits are 'wasted'). Generally speaking, the more patterns that a message contains, the more redundancy there is built into it, and the lower the total information transfer rate. Of course, redundancy may be desirable, and is usually deliberately built into human messages, because the transmission process may introduce errors. But the optimal data transmission rate is one that has *no* patterns whatsoever, and is therefore random. Randomness does not mean nonsense. If one has the key to decode the message, the information is optimally packaged. Without the key, however, the message would just come across as a form of noise.

There is an obvious tension between being conspicuous and optimal data packaging. Noise in a radio telescope may not present itself to us as an intelligent signal. We are surrounded by random noise – from quantum fluctuations in atomic systems to the hiss from the sky produced by the primordial cosmic microwave background radiation. Would we know whether some of the cacophony of the universe is in fact optimally encoded messages from distant civilizations, and not natural scrambling? The short answer is that, without the code, we wouldn't know. We could be in the midst of a gargantuan alien data exchange, and blissfully unaware of it. In *Contact*, Sagan had the aliens send a sequence of prime numbers as the 'Hi, guys!' part of their message to attract attention. To a mathematician, prime numbers are not random. To take a humbler example, smoke rising haphazardly from a hillside might be either a natural bushfire or a campfire, but a patterned sequence of discrete smoky puffs would indicate that a campfire is being used to send a signal. The same principle applies to a lighthouse or any other beacon. So the 'hook' part of an alien signal intended for strangers should be conspicuously *non*-random, but the content of an information exchange between consenting radio pals would most likely be random (assuming the aliens care about transmission efficiency). For an astronomer to twig on that a source is artificial, it would need some sort of signature of intelligence or technology. If the signal is not directed at us specifically then it may lack any attention-grabbing hook, but other features might give the game away. For example, if the signal was bright enough to rise above the background noise, was narrow band in frequency, and emanated from a nearby star with a known Earth-like planet, we would definitely take notice.

Suppose, then, astronomers pick up a signal that looks artificial in some way, but lacks any indication that it is either intended for humanity specifically, or is being broadcast for general cosmic consumption (as in the case of a beacon). In terms of an official statement, the situation would be little different from the scenario I considered in the previous section, and the discovery should be made public in the conventional manner. So let me move on to the least likely, but easily the most momentous, scenario: the receipt of a message deliberately crafted for mankind.

SECRETS FROM THE STARS

If an alien civilization were to send us a customized message then all bets are off. Right from the outset some extremely hard choices would need to be made, choices that the Post-Detection Taskgroup has pondered. The first decision would be whom to tell and how. In this scenario, the published Protocol would almost certainly break down. I personally feel that the implications of simply receiving such a message would be *so* startling and *so* disruptive that, although eventual disclosure is essential, every effort should be made to delay a public announcement until a thorough evaluation of the content had been conducted, and the full consequences of releasing the news carefully assessed in light of the Taskgroup's recommendations. Ideally, information about the astronomical coordinates of the transmitter should be restricted to the astronomers involved, for reasons I will come to shortly. As we've seen, however, keeping the lid on such a discovery would present enormous obstacles. Even governments – which have so far shown little interest in SETI – would presumably at last take notice, and no doubt would also want to take charge. In my view, however, the less government involvement at the evaluation stage, the better. Any attempt to control as opposed to facilitate the scientific assessment would in all probability be counter-productive.

The way in which events unfold would depend on the actual content of the message. Foremost is the question of decoding it. Presumably ET won't speak English, or any other human language, unless the alien intelligence has been monitoring our broadcasts. By common consent, mathematics, being culturally neutral and forming the basis of the universal laws of nature, would be the lingua franca of interstellar discourse. Sagan's *Contact* had a message in the form of pictures, with prime numbers used to structure a pixilated array. Remember that this will be a one-way communication from a truly alien species, not a real-time dialogue with smiles, frowns, finger-pointing and other gestures that humans use to get their meaning across even to total strangers. The aliens can share with us more than just mathematics, however. There is cosmography too. We live in the same universe and very likely the same galactic neck of the woods, so symbols to denote stars and other astronomical objects would be readily understood by us. By extension, ideas about shared basic science could be communicated in pictures and correlated with symbols. Bit by bit we might build up more abstract notions and begin to learn their language. Obviously this makes huge assumptions about the mental architecture of an alien mind. The very notion of language and its symbolic representation has emerged from the study of human beings. Who can say whether aliens would think or attempt to communicate in the same way?

It would be a huge undertaking to make sense of the message, hampered by the fact that it might be incomplete or distorted by noise. Decoding it could take a very long time, perhaps involving years of meticulous work and computer analysis before we had any idea of what we were dealing with. I cannot imagine how the scientists involved would be left to work in peace to do this. Nevertheless, a drawn-out process of analysis would do much to reduce the cultural shock that would follow the initial announcement. As Sagan expressed it, 'the decoding of the message, the understanding of the contents, and the extremely cautious application of what we are taught might take decades or even centuries . . . A message that will take a long time to decode and understand will not be very . . . disorienting to the average man.'[13]

Let us assume that, sooner or later, the gist of the content begins to emerge. What then? Now we really are in guesswork territory. What would ET want to say to us? The simplest message would be along the lines of 'We are here and you are there, and we just called to say hello.' More thought-provoking would be 'We invite you to join the Galactic Club and exchange information with your cosmic neighbours.' We can also imagine communications with alarming content, such as 'Your civilization is in grave danger. We have spotted a giant comet heading your way.' Then there are moral missives: 'Our instruments have detected nuclear explosions on your planet and we strongly advise you to sort out your problems – previous civilizations we know that have exploded nuclear weapons didn't survive long.' This last one is unlikely to come soon, given that information about the first nuclear explosion has reached less than seventy light years into space. Evidence for the early build-up of human-generated carbon dioxide would have penetrated farther, however. Maybe that would elicit a warning along the lines of 'Stop burning fossil fuels, you foolish beings.'

Harder to fathom is the impact of a message that imparts important scientific or technological information. Most worrying of all would be one that merely handed us on a plate a revolutionary item of technology, e.g. a new source of energy, or a technique for engineering designer life forms reliably. The problem here is that the group that possessed the knowledge first would be in a position of incomparable power. Nations, scientific organizations, companies and other special-interest groups would fight tooth and nail to gain access to, and control over, gems of alien know-how. Outright warfare might follow the scramble to grab the information. One can only hope that the aliens would recognize the dangers and refrain from handing out scientific secrets like sweets.

A less risky way for a benevolent alien civilization to offer technological help would be to issue an invitation for us to download scientific data at some point in the future, subject to safeguards and provisions to avert an unseemly squabble over who gets first peek, plus some clear assurances about how we would use the information afterwards. For example, a longstanding hope for solving the world's energy crisis is controlled nuclear fusion – the process that powers the sun. Experiments were begun in the 1950s, with the expectation that fusion power would be a commercial reality within thirty years. Today, experiments with nuclear fusion continue, but the promise of unlimited cheap energy remains a distant dream. The main technical obstacle is finding a way to confine the ultra-hot hydrogen gas, which has a tendency to become unstable (this process is hot fusion, not the dubious 'cold fusion' I discussed on p.173). A helpful tip from ET could enable scientists to solve the stability problems. However, the sudden transformation of our industry to almost-free fusion power would seriously rock the economic boat and change the geopolitical landscape overnight. Forward planning of some decades would be highly advisable.

IMPACT ON SCIENCE, PHILOSOPHY AND POLITICS

The mere knowledge that another technological community exists would imply that there are, have been and will be very many such communities; the probability that there are two, but only two, civilizations in the galaxy is very low. We could straightaway conclude that f_l and f_i in the Drake equation are not, after all, close to zero. The hunt would then begin in earnest for other alien civilizations, possibly closer, and a serious attempt would be made to find alien artifacts on or near Earth. Astrobiology as a whole would receive a massive fillip, because to know that f_l is not a tiny number means we can expect to find at least microbial life in many Earth-like settings, perhaps even within our own solar system.

There would also be a major paradigm shift among scientists. According to the orthodox scientific world view the great sweep of cosmological history is organized around two fundamental principles: the Copernican principle and the second law of thermodynamics. The latter, which I touched on in Chapter 7, concerns the unremitting rise in entropy in all physical systems and the resulting one-way slide of the universe from order to chaos, tending towards what physicists call its 'heat death'. The most conspicuous manifestation of the second law at work is the way stars eventually exhaust their stock of nuclear fuel and burn out. In the very far future, not just starlight, but all forms of useful energy, will be completely dissipated. To a thermodynamicist, the history of the universe is one of inexorable degeneration and decay. 'We are the children of chaos,' writes the chemist Peter Atkins, 'and the deep structure of change is decay. At root, there is only corruption, and the unstemmable tide of chaos. Gone is purpose; all that is left is direction. This is the bleakness we have to accept as we peer deeply and dispassionately into the heart of the Universe.'[14]

Viewed through the eyes of a cosmologist, however, the same facts could take on a different hue. The universe began in a rather bland state – a hot uniform soup of subatomic particles. Over time, through a sequence of self-organizing processes, it has increased enormously in richness and complexity. Matter aggregated into galaxies, which then differentiated into stars. Heavy elements were made, leading to the formation of planets. Planets produced rocks and clouds and hurricanes and, in at least one case, life. Starting with a handful of humble microbes, life on Earth has diversified over billions of years into the astonishing variety of elaborate forms we see today. A cosmologist might prefer to describe the history of the universe as one of continual enrichment rather than relentless degeneration and decay. However, the two accounts – thermodynamic and cosmological – are not contradictory. They simply emphasize different aspects of change. They are consistent because every self-organizing process, every new species of life, comes with a thermodynamic price in the form of increased entropy, which serves to hasten the cosmic heat death.

Now we reach the point I want to make. There is a strong temptation to describe the cumulative enrichment of the universe as 'progressive'. It looks as if there is some sort of overarching principle at work – a principle of advancing complexity and organization – which applies to everything from the formation of galaxies to the evolution of multicelled life. Onward and upward the march seems to go – to brains, cognition, intelligence and technological society. SETI sits at the pinnacle of that hypothesized swoop, predicated on the assumption that there is indeed a principle of advancing complexity, playing out across the galaxy and the wider universe, facilitating the emergence of life, intelligence and technology wherever they have an opportunity to flourish. It is an inspiring vision. But is it credible? The majority of scientists would say no, dismissing such ideas as quasi-religious. In Chapter 4, I explained how the notion of 'progress' is a highly contentious and sensitive issue among biologists. It rests uncomfortably within the reigning paradigm of Darwinism, which rejects any suggestion that nature can 'look ahead' and legislate a systematic overall directionality in evolution. As for physics and chemistry, decades of research into complex systems have so far failed to unearth any general 'law of progress', only vague trends and specific examples involving special circumstances. The discovery of alien technology would settle this matter in short order, and demonstrate, against the prevailing orthodox scientific sentiment, that the cosmos is indeed subject to some sort of universal principle of advancing organized complexity.[15]

The impact on philosophy would be equally profound. The thermodynamic view of nature, in stressing the remorseless decay and impermanence of all physical systems, has long bolstered a nihilistic philosophy, or at best stoic acquiescence, in the face of a pointless, aimless universe enduring a lingering heat death. A century ago the hugely influential British philosopher Bertrand Russell wrote gloomily about the 'unyielding despair' it invites one to accept when contemplating 'the vast death of the solar system'.[16] The contrasting view – that the universe is pregnant with hope and potential, and is riding an escalator of growth to glories new – underpinned the countervailing visions of progress towards Utopia espoused by Russell's Continental contemporaries,[17] which contributed to the rise of European socialist thought. The same divergence of opinion prevails today. Mankind in the twenty-first century faces an uncertain future, and many distinguished scientists are pessimistic that we have any long-term future at all.[18] Yet set against this are predictions of accelerating technological progress, promising the elimination of all society's ills, as expounded for example by Freeman Dyson[19] and the futurist Ray Kurzweil.[20]

The knowledge that an alien community had endured for eons and overcome the multiple problems that mankind currently faces would rekindle human Utopian dreams and become a strong unifying force on our planet. To glimpse a trajectory of human progress mirrored in the stars would have a galvanizing effect far greater than any political rhetoric. In our present state of ignorance it is possible to believe either account of the future: pessimistic or optimistic. But to know we are not the only sentient beings in a mysterious and sometimes frightening universe would provide a dramatic message of hope for mankind.

IMPACT ON RELIGION

Undoubtedly the most immediate impact of an alien message would be to shake up the world's faiths. The discovery of *any* sign that we are not alone in the universe could prove deeply problematic for the main organized religions, which were founded in the pre-scientific era and are based on a view of the cosmos that belongs to a bygone age. Although the cosmological discoveries of Copernicus, Galileo, Einstein and Hubble proved discomforting for religion, they were eventually accommodated because most religions make no serious attempt to describe the physical universe in a scientific manner. Their creation myths are poetical and symbolic, rather than factual. Two thousand years ago, few people had any inkling that a vast universe lay beyond the sky: Earth's surface and its life *were* creation. The reason that scientific cosmology, with its billions of galaxies scattered across the chasms of space, failed to demolish established religion is because religious faith is primarily concerned with *people*, not the universe. Indeed, most religions focus on one particular species that has existed on one planet in one galaxy for a mere one hundred thousandth of the age of the universe, a species that nevertheless is said to enjoy a special relationship with the very Architect of the cosmos. The danger posed by SETI is that religion primarily concerns not the vastness and majesty of the cosmos, but *the affairs of sentient beings*.

Christianity is the religion most challenged by the concept of extra-terrestrial beings, because Christians believe that God became a human being (specifically, a Jewish political dissident). Jesus Christ is called the Saviour precisely because he took on human flesh to save human-kind. He did not come to save the whales or the dolphins or the gorillas or the chimpanzees, or even the Neanderthals, however noble or deserving those creatures may be (or were). Jesus Christ was the saviour of *Homo sapiens*, specifically: one planet and one species. The plausibility of such an extraordinarily focused divine mission was much easier to accept when most people believed – as they did two millennia ago – that there was only one Earth and one intelligent species, when nothing was known of the now vanished Neanderthals, and little thought had been given to the possibility of alien beings on other worlds.

The problem for Christianity is thrown into sharp relief when account is taken of the relative state of advancement of alien civilizations. As I have stressed, if intelligence is widespread in the universe, there will be communities of beings who may have reached our stage of development millions of years ago. Those beings are likely to be far ahead of us not only scientifically and technologically, but ethically too. Quite possibly they will have used genetic engineering to eliminate grossly criminal or antisocial behaviour. By our standards they would be truly saintly.[21] And herein lies the real crisis for Christianity. If we miserable humans get to be saved, surely the saintly aliens deserve a chance too?

Well, what does the Church have to say on the matter? The problem of extraterrestrial life, while hardly a Premier League issue, has not been totally ignored by theologians. A search of the literature reveals two escape clauses whereby aliens could be saved. The first appeals to multiple incarnations: one saviour for each deserving species – 'God taking on little green flesh to save little green men' was the refreshingly blunt way an Anglican priest once expressed it to me. The problem with this idea is that the incarnation (meaning 'God becoming flesh') is supposed to be a unique event: the Bible says that Jesus is God's *only* begotten son. Incarnations on billions of planets is regarded as a heresy by many Christians. The other resolution is to suppose that there is only one incarnation and one saviour, in the form of the terrestrial Jesus Christ, and that it is the God-given destiny of mankind to 'spread the word' around the universe. Humans thus assume the responsibility for a sort of cosmic crusade, presumably at first by radio, raising the amusing prospect that if we ever make contact with ET, Christians may present themselves as the aliens' route to salvation rather than vice versa![22]

Both the above-mentioned scenarios have been mulled over by theologians, usually with the reassuring conclusion that ET is in fact no threat to Christianity. Consider, for example, the recent statement by the Reverend José Gabriel Funes, head of the Vatican Observatory and a scientific adviser to Pope Benedict XVI, who is distinctly sanguine about extraterrestrial intelligence. 'How can we exclude that life has developed elsewhere?' he remarked in a newspaper interview. 'Just as there is a multiplicity of creatures on earth, there can be other beings, even intelligent, created by God.' But is Christianity thereby imperilled? Not at all, according to Fr Funes: 'The extraterrestrial is my brother.'[23]

Shortly after this comment was made, a survey was published in which 1,135 people of several faiths were asked whether the discovery of extraterrestrial intelligence would have a negative impact on specific religions. The study was conducted by the Lutheran theologian Ted Peters, who has a longstanding interest in the theological implications of aliens.[24] Remarkably, very few religious adherents thought there was a problem. Most said their faith could readily accommodate the existence of advanced alien beings without too much disruption to their core beliefs. Many respondents, echoing Fr Funes, even welcomed the idea of ET, and thought it painted a richer picture of God's creation. However, most of the comments had an air of sweeping the problem under the carpet. Very few of the Christian respondents tackled the theological minefield of the uniqueness of the incarnation and the species-specific nature of salvation. A handful did identify the conundrum, but no novel solutions were proffered.

Christians haven't always been so laid-back about the matter. When Bruno proposed that there were many inhabited worlds, he was condemned to death in 1600 for heresy.[25] Bruno's dreadful fate did little to dampen enthusiasm for debate about extraterrestrial life, and belief in alien beings became widespread in Christian Europe. But the stubborn problem of the incarnation always lurked in the background. William Whewell was an early-nineteenth-century Cambridge University philosopher, famous for coining the term 'scientist', and, like Isaac Newton before him, was Master of Trinity College. His academic position held the grand title of Professor of Moral Theology and Casuistical Divinity. Articulating the prevailing view, Whewell initially argued in favour of extraterrestrial beings, but by 1850 doubts began to creep in, fuelled precisely by theological concerns about the incarnation and the salvation of mankind. In an unpublished manuscript entitled *Astronomy and Religion* he wrote:

God has interposed in the history of mankind in a special and personal manner . . . what are we to suppose concerning the other worlds which science discloses to us? Is there a like scheme of salvation provided for all of them? Our view of the saviour of man will not allow us to suppose that there can be more than one saviour. And the saviour coming as a man to men is so essential a part of the scheme . . . that to endeavour to transfer it to other worlds and to imagine there something analogous as existing, is more repugnant to our feeling than to imagine those other worlds not to be provided with any divine scheme of salvation . . .[26]

In other words, said Whewell, there are no extraterrestrials worthy of being saved. His stern deliberations culminated in a book, published anonymously in 1854, entitled *Of the Plurality of Worlds*, in which he attempted to deploy scientific arguments to bolster what was primarily a Christian objection to the existence of aliens.[27]

Nevertheless, the contrary view – that there are countless planets hosting positively saintly beings – has proved popular among Christians too. In 1758, Emanuel Swedenborg, a Swedish scientist, philosopher and mystic, who still commands a cult following today, offered a way out of the theological quagmire in a curious little book entitled *Earths in the Universe*.[28] Like many eighteenth-century scholars, Swedenborg was convinced – also on theological grounds! – that other planets, including those in our solar system, were inhabited. He even went as far as to describe the appearance, clothing, family structure, religious practices, houses and other mundane aspects of the aliens' lives, information he claimed to have accessed through mystical revelation. Some alien societies, Swedenborg declared, were positively idyllic. On Mars, for example, the inhabitants were of a much friendlier disposition than earthlings; when strangers meet 'they are instantly friends.' Furthermore, 'everyone there lives content with his own goods', and precautions are taken against 'the lust of gain' lest anyone 'should deprive others of their goods'.[29] In spite of this alleged Martian Utopia, Swedenborg insisted that Earth alone hosted an incarnation. His chapter 'The reasons why the Lord willed to be born on our Earth, and not on any other' explains his reasoning. God selected Earth in order to deliver 'the Word . . . the Divine Truth', with the express purpose that it should first be communicated across our planet, and then passed to other planets.[30] How, you might wonder? Lacking knowledge about the possibility of radio, Swedenborg invoked 'spirits and angels' as the mode of communication to the extraterrestrials. On the problem of the species-specific nature of the incarnation, Swedenborg had a quaint solution. The extraterrestrials were, he said, *humans too*: 'there are earths in immense numbers, inhabited by human beings, not only in this solar system, but in the starry heaven beyond it.'[31] Thus, when Jesus Christ died to save mankind, the definition conveniently extended to embrace the aliens.

Swedenborg's concept of a theologically privileged Earth, with 'the Word' spreading out into space like ripples from a stone thrown into a pond, was adopted in the twentieth century by none other than E. A. Milne, a British mathematical physicist and cosmologist of some distinction, who was a professor at Oxford University. In his book *Modern Cosmology and the Christian Idea of God*, published in 1952, Milne wrote:

> God's most notable intervention in the actual historical process, according to the Christian outlook, was the Incarnation. Was this a unique event, or has it been re-enacted on each of the countless number of planets? The Christian would recoil in horror from such a conclusion. We cannot imagine the Son of God suffering vicariously on each of a myriad of planets. The Christian would avoid this conclusion by the definite supposition that our planet is in fact unique. What then of the possible denizens of other planets, if the Incarnation occurred only on our own?[32]

Quite. Milne got it precisely. He went on to suggest that the theological problem would be circumvented if the Word could be spread from Earth using radio telescopes, which is at least an improvement on the 'spirits and angels' of Swedenborg.[33]

It will be evident from these selected quotations that Christian theology is in a frightful muddle when it comes to extraterrestrial beings, and that a positive result from SETI would immediately open up a horrible can of worms, whatever bland assurances have been given by religious leaders so far.[34] In fact, I would go so far as to say that the discovery of aliens would deal a severe blow not only to Christianity, but to all mainstream religions. I am not saying that what we may loosely call the spiritual dimension of human life would be eclipsed or belief in some sort of wider meaning or purpose in the universe negated. Buddhists would doubtless continue to seek the path of enlightenment through inner reflection, even when armed with the knowledge of intelligent life beyond Earth. What *is* clear, however, is that any theology with an insistence on human uniqueness would be doomed. How this would actually play out in terms of social and political disruption across the world is difficult to predict. Although slow to change, religion is very adaptable. Over the centuries it has managed to come to terms with Copernican cosmology, Darwinian evolution, genome sequencing and other unsettling scientific developments. Of these, evolution was the hardest to swallow, because of its implied threat to the unique status of *Homo sapiens*. The discovery of advanced extraterrestrial beings would represent a far more explicit threat of the same nature, and prove that much harder to assimilate.

OF GODS AND MEN. IS SETI ITSELF A RELIGION?

Humans have a basic need to perceive themselves as part of a grand scheme, of a natural order that has a deeper significance and greater endurance than the petty affairs of daily life. The incongruous mismatch between the futility of the human condition and the brooding majesty of the cosmos compels people to seek a transcendent meaning to underpin their fragile existence. For thousands of years this broader context was provided by tribal mythology and storytelling. The transporting qualities of those narratives gave human beings a crucial spiritual anchor. All cultures lay claim to haunting myths of otherworldliness: from the Dreaming of the Australian Aborigines to the *Chronicles of Narnia*, from the Nirvana of Buddhism to the Christian Kingdom of Heaven. Over time, the humble campfire stories morphed into the splendour and ritual of organized religion and the great works of drama and literature. Even in our secular age, where many societies have evolved to a post-religious phase, people still have unfulfilled spiritual yearnings. A project with the scope and profundity of SETI cannot be divorced from this wider cultural context, for it too offers us the vision of a world transformed, and holds the compelling promise that this could happen any day soon. As the writer David Brin has pointed out, 'contact with advanced alien civilizations may carry much the same transcendental or hopeful significance as any more traditional notion of "salvation from above".'[35] I have argued that if we did make contact with an advanced extraterrestrial community, the entities with which we would be dealing would approach godlike status in our eyes. Certainly they would be more godlike than human-like; indeed, their powers would be greater than those attributed to most gods in human history.

So is SETI itself in danger of becoming a latter-day religion? The science fiction writer Michael Crichton thought so. 'SETI is unquestionably a religion,' he said bluntly, in a 2003 speech at the California Institute of Technology.[36] Crichton was objecting to the widespread use of the Drake equation when many of the terms it includes are pure guesses. 'Faith is defined as the firm belief in something for which there is no proof,' he explained. 'The belief that there are other life forms in the universe is a matter of faith. There is not a single shred of evidence for any other life forms, and in forty years of searching, none has been discovered. There is absolutely no evidentiary reason to maintain this belief.' In similar vein, George Basalla, a University of Delaware historian, argues that doggedly pursuing contact with aliens in the face of fifty years of silence betrays a kind of religious fervour, bolstered by a vestige of the belief that the heavens are populated by superior beings.[37] The writer Margaret Wertheim has studied how the concept of space and its inhabitants has evolved over several centuries. She traces the modern notion of aliens to Renaissance writers such as the Roman Catholic Cardinal Nicholas of Cusa (1401–64), who considered the status of man in the universe in relation to celestial beings such as angels. 'Historically, this may be seen as the first step in a process that would culminate in the modern idea of aliens,' writes Wertheim. 'What are ET and his ilk, after all, if not incarnated angels – beings from the stars made manifest in flesh?'[38]

With the arrival of the scientific age, speculations about alien beings passed from theologians to science fiction writers, but the spiritual dimension remained just below the surface. Occasionally it is made explicit, as in Olaf Stapledon's *Star Maker*, David Lindsay's *A Voyage to Arcturus*, or Steven Spielberg's *Close Encounters of the Third Kind*, which is strongly reminiscent of John Bunyan's *A Pilgim's Progress*.[39] These are iconic images that resonate deeply with the human psyche, and shadow the scientific quest to discover intelligent life beyond Earth. Most SETI astronomers vehemently reject the claim that there is a religious dimension to their work; they regard the existence of aliens as simply a hypothesis to be tested. But for many non-scientists, the fascination of SETI is precisely its quasi-religious quality, and its tantalizing promise of celestial wisdom and unbounded riches in the sky – just a radio signal away.

10

Who Speaks for Earth?

Take me to your leader!

Plea of a thousand cartoon aliens

SHOUTING AT THE HEAVENS

Imagine that the day has arrived. Humanity has received a message from an alien civilization, directed at Earth. The message has been decoded and the aliens are asking for contact. Should we respond? If so, what do we say? Above all, who speaks for Earth?

The SETI Post-Detection Taskgroup has already begun to wrestle with these thorny problems, for the simple reason that some people have jumped the gun and begun transmitting messages anyway, a practice known as active SETI or METI (Messaging to Extraterrestrial Intelligence). Radio METI began in earnest in 1974, when the Arecibo radio telescope was employed to transmit a message to the M13 globular cluster of stars 25,000 light years away. A more recent attempt was made in 2009 when a large radio telescope in Ukraine was used to beam fifty photos, drawings and text messages at the planet Gleise 581C, located twenty light years away. The target is one of a handful of newly discovered extra-solar planets thought to be capable of supporting life.

Some people are implacably opposed to METI on the grounds that broadcasting willy-nilly into space, deliberately attracting attention to ourselves, is reckless. An obvious fear is that advertising the existence of our wonderful life-supporting planet might invite an alien invasion. A leading critic of METI is the writer and commentator David Brin, who coined the phrase 'shouting at the cosmos'. He is dismayed by the happy-go-lucky attitude of a new generation of SETI fans, especially those from the former Soviet Union, who advocate greatly expanding the METI programme in an ad hoc manner without much forethought or attempt at debating the issue. And it's true that METI attracts far more attention than SETI, primarily because something actually happens – a message is sent! By contrast, all SETI astronomers do is passively listen. METI is popular with young people when the content of the message is opened up to the public; the recent Ukraine transmission followed a competition launched via a social networking site called Bebo, which boasts 12 million users. Brin's position is that prudence should prevail over popularity. He has called for an international protocol that asks for all of those people controlling radio telescopes to '*forbear from significantly increasing Earth's visibility* with deliberate skyward emanations, until their plans were first discussed before open and widely accepted international fora [his italics]'.[1] His sentiments have been strongly endorsed by David Whitehouse. 'If we don't know what's out there,' writes Whitehouse, 'why on Earth are we deliberately beaming messages into space, to try and contact these civilizations about which we know precisely nothing?'[2]

Champions of METI, such as Alexander Zaitsev of the Russian Academy of Sciences, dismiss Brin's concerns. They point out that we are already broadcasting. Our radio and television programmes are sweeping across the galaxy at the speed of light: we can't get them back. A sufficiently sensitive antenna could detect them, and our cover would be blown. However, as I mentioned earlier, our TV transmissions are actually exceedingly weak. Military radars pack much more punch, as do the occasional radar pulses directed at planets and asteroids for scientific purposes. But these beams are sporadic and narrow; ET could easily miss them. So all in all, there is a good chance we have so far escaped detection (by radio at least) even if the galaxy has legions of alien civilizations armed with enormous radio antennas. No doubt this debate will rage for a while yet, but it seems to me largely irrelevant, because whatever scientists and commentators may think, the reality is that a motivated millionaire can build a radio telescope and blast the heavens to his heart's content, and there's very little anyone can do about it.[3] METI cannot realistically be policed – at least, no international agency able to do so has the slightest interest in the subject one way or the other.

I am clear in my own mind that the danger from METI is minuscule. Fear of the unknown is understandable, but if we always wait until we are sure there are no demons lurking in the dark we would never do any science and never explore our world. Prudence is wise, but prudence should not mean paralysis. We need to ask why aliens would be interested in harming us or invading. If Earth is attractive as a potential alien habitat, the aliens will know this already without our help. Evidence for oxygen, water and plant life can be obtained spectroscopically from a great distance, even with foreseeable human technology. So we are right back to the Fermi paradox: if they were going to come here for our planet – as opposed to us – they would have showed up long ago. In any case, our radio messages are irrelevant if the planet is what they want. The only additional information to be gleaned from radio communications is that Earth also hosts intelligent life capable of building radio transmitters. Some people worry about enslavement, but that is foolish. A technological community advanced enough for interstellar travel is hardly going to have a labour shortage. It could more easily build robots or bio-machines to do the necessary grunt work. We might conceivably be regarded as a cultural resource or a biological curiosity, and therefore worth preserving. If so, there is no danger. The concern I voiced in Chapter 8, that humans might be duped into building a hostile alien from genetic instructions, is not relevant to METI. That scenario would need careful consideration only if we receive a meaningful message from *them*.

The greatest danger to humanity is if a nearby alien community judges us to be a threat. Given our warlike history, that is not an unreasonable conclusion. The aliens might decide to mount a preemptive strike for the greater good of the wider galactic community. And could we blame them, given that some of our own governments have used precisely the same logic against perceived terrestrial enemies? If twenty-first-century human democracy is anything to go by, it may require no more than a thin pretext for extraterrestrials to 'take out our weapons of mass destruction'. But even if this gloomy assessment is correct, METI would not increase the risk of bringing fire and brimstone down on us. In fact, it may serve a useful purpose if we could signal our best intentions to ET, in spite of our penchant for warmongering at home. Just how we could convince aliens that we wouldn't try to blow them away with our missiles and nuclear warheads is another matter. Such a message would in any case be a lie. Humans have fought each other for millennia over tiny differences in race, religion or culture. Imagine how most people would react to beings that were *truly* alien – not only a different species, but a different life form altogether, with unknown motives and non-human feelings. Fear and revulsion could well provoke a shoot-first-ask-questions-later response. My personal message to ET is to 'Keep well clear and defend yourself', before stepping into the hornets' nest of our militaristic society. I hope such a warning would be regarded in itself as sufficiently altruistic to avert a pre-emptive strike.

I am in favour of METI, not just because I think there isn't a snowball's chance in hell of anyone out there picking up the signals, but because the act of designing and transmitting messages to the stars serves many noble purposes, such as raising interest in science in general and SETI in particular, and in encouraging people – especially young people – to think about the significance of humanity and the vastness of the universe, and to reflect on the common factors among our disparate cultures that we wish to preserve for posterity. METI is good for humanity and almost certainly completely harmless, given the infinitesimal chance that randomly beamed signals will ever be detected by a malevolent alien civilization.

WHAT SHOULD WE SAY?

In the present context, METI is little more than a harmless stunt. The situation would be dramatically different, however, if we had actually located an extraterrestrial civilization. In that case, it is essential that wiser counsels prevail. Item 7 of the IAA's 'Declaration of Principles Concerning Activities Following the Detection of Extraterrestrial Intelligence' enshrines the need for caution:

> No transmission in response to a signal or other evidence of extraterrestrial intelligence should be sent until appropriate international consultations have taken place.[4]

Unfortunately history gives me very little confidence in the efficacy of 'international consultations'.

On the matter of who gets to respond 'officially', I foresee all sorts of problems. A message concocted by a committee would be a recipe for the lowest common denominator, and is likely to consist of banalities. A statement solely by a politician or religious leader is too horrible to contemplate. A potpourri of comments, where each cultural group has its say in the interests of equity or democracy would most likely be judged an incoherent muddle. This sort of pointless gimmick was tried in 1977, when the two Voyager spacecraft, which are going fast enough to leave the solar system, carried identical phonographs. The records convey greetings in fifty-five languages, bird and animal sounds, a selection of music ranging from string quartets to rock and roll, and sombre written statements from President Jimmy Carter and UN Secretary-General Kurt Waldheim. If ever aliens were to chance upon Voyager drifting in interstellar space, I dread to think what they would make of it all.

Could scientists improve on this? On my office wall hangs a fine plaque, presented to me by NASA. It is a replica of the ones conveyed aboard the spacecraft Pioneer 10 and 11 (Fig. 13). Pioneer 10 was the first manmade object to leave the solar system, so NASA thought it would be a nice, albeit futile, gesture to convey a message to aliens. As a symbolic act, it is a great idea, and I am proud to possess a replica. My beef is not with the gesture itself, but the content. The plaque was designed by Carl Sagan, Linda Salzman Sagan and Frank Drake, and shows a picture of a male and female form, one with a hand raised in greeting, together with an image of the spacecraft and some technical data. A line symbolizes the trajectory of the spacecraft showing it originating on the third planet from the sun. Our galactic coordinates are encoded in a clever way, by showing the locations and frequencies of a set of pulsars, from which the sun's position in the galaxy could be reconstructed by a distant civilization using elementary geometry.

This plaque may be worthless as far as signalling the aliens is concerned, but it speaks volumes about humans. A brief message to an unknown alien community should presumably reflect the things that we consider most significant about ourselves. The picture is dominated by the human shapes, yet our physical form is probably the *least*

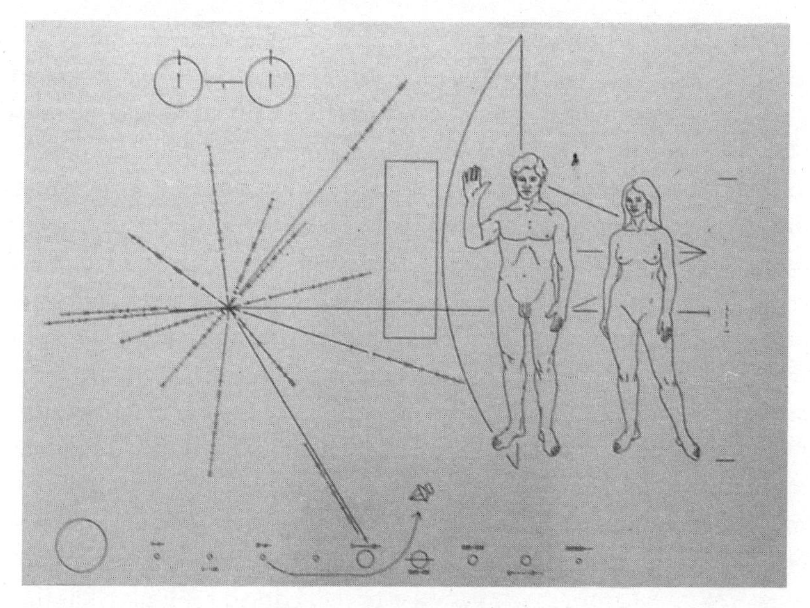

Fig. 13. Pioneer plaque.

significant thing we can say. It is almost completely irrelevant both scientifically and culturally. To put it bluntly, who gives a damn what we *look* like?[5] The raised hand part is the height of absurdity: such a culturally specific mannerism would be utterly incomprehensible to another species, especially one that might not have limbs. Showing the provenance of the spacecraft within the solar system is of little relevance. If the sun's location is established, it wouldn't take a genius to figure out which planet had intelligent life. The plaque also conveys the information that humans are carbon-based. But we hardly need to teach ET chemistry and biology. Carbon is probably the only life-giving element, but if the aliens really wanted to know, they could scour the spacecraft for remnants of terrestrial microbes. Thirdly, and more seriously, a preoccupation with what we are made of is almost as parochial as concern over our physical form. Surely the essence of humanity is what we do and think, not the chemical make-up of our bodies.[6]

This half-hearted attempt to put our stamp on the cosmic community is distinctive in its narrow-mindedness and preoccupation with twentieth-century science and human affairs. In fact, it addresses the sort of topics that appear on the agenda of SETI conferences, but are exceedingly unlikely to be on the agendas of conferences in a 10-million-year-old alien civilization, especially one in which machines/computers are doing the intellectual heavy lifting. As calling cards they are effectively useless.

Well, can I come up with anything better? I hope so. One way to approach the topic is to imagine that our species is about to be annihilated, and we wish to leave a record of our erstwhile existence, perhaps for a future intelligent species that may evolve on Earth in the fullness of time. What would we want to say about ourselves? What do we most value? Which products of our culture are quintessentially human? We might take great pride in our technological accomplishments, such as the Moon landings, or particle accelerators, or genome sequencing; but then again, maybe not. My grandmother's response to the Apollo programme was 'Why do they want to go to the Moon?' She couldn't see the point. In the grand cosmic scheme of things, technological products may cut little ice, especially among a species that has no left-brain/right-brain dichotomy, no art–science schism.

When it comes to cultural achievements, we are in even murkier waters. Religion I have already dealt with: most religions are highly geocentric and anthropocentric (even ethnocentric), deeply rooted in evolutionary psychology and recent human history. They would be totally meaningless to an alien mind. Great works of literature or poetry are equally parochial, because they celebrate and analyse the realm of human affairs and relationships. Art may have more widespread appeal, although beauty is very much in the eye of the beholder. It is not inconceivable that there are universal aesthetic principles, having to do with symmetry for example.[7] Even an alien mind may recognize certain forms of visual art to be making a statement to which it could relate in a general sort of way. But there is no accepted theory of art that isn't intimately tied to the human cognitive system. The same goes for music and humour: they work well for humans because we share most of our neural architecture. An alien brain will be wired differently, so aliens will find different things pleasing, things that are probably incomprehensible to us. I have left out sport, economics and stamp-collecting for reasons that hardly need to be spelled out.

In the trade-off between content and comprehensibility, we would be wise to err on the side of the latter. There is little point in sending obscure philosophical thoughts about emergence, post-modernism or moral relativism without a library of definitions and background information. Even biology is problematic: apart from the principle of Darwinian evolution, we don't really know any universal biological laws, so communicating details of protein assembly or gene networks might be of little value. (That may change as our understanding of bio-systems improves.)

Which leaves us with mathematics and physics. The deepest products of the human mind are arguably the mathematical theorems that have been constructed by some of the world's most brilliant thinkers. Gödel's incompleteness theorem, for example, is so profound that it is possible that no theorem in the universe can trump it.[8] (I make this bold claim because Gödel's theorem is a very general statement about what *cannot* be known or proved – ever, in principle – rather than about something specific which is known.) Mathematics occupies an unusual place in our culture in that it is a product of the human mind, and yet it transcends the mind. Any sufficiently advanced being elsewhere in the universe could prove the same theorems based on the same logical principles. Given that the universal laws of physics are manifested in the form of elegant mathematical regularities, it is clear that mathematics is the key to bridging the gulf between human and alien cultures. If aliens know any science, or have developed any advanced technology at all, then they will be familiar with mathematics. They will even be familiar with the *same* mathematics as we know. To take an example, Maxwell's laws of electromagnetism are observed to apply everywhere in the universe, so if the aliens understand the principles of radio – which we are assuming is a prerequisite for radio contact at least – then they will know Maxwell's equations. What else? Einstein's general theory of relativity has been described as the pinnacle of human intellectual achievement – it is certainly an impressive accomplishment. Then there is the quantum theory of fields and other esoteric products of theoretical physics that accord well with experiment. If the aliens have gone beyond radio, they will presumably know where the general theory of relativity and quantum field theory fit into the sum total of knowledge about the universe. If we inform them that we have attained that degree of understanding, it will be a benchmark of sorts for them to judge our level of advancement.

The reader might be thinking, 'Well, he would say that, wouldn't he? It's just what you'd expect from a theoretical physicist. Davies is as parochial as the rest of us.' But let me defend my choice. Part of the reason I became a theoretical physicist is precisely because mathematics and physics have universal significance. I was attracted to these subjects because they do seem to transcend human affairs, to put us in touch with the deepest workings of nature. If, wearing my hat as Chair of the SETI Post-Detection Taskgroup, I get to reply to ET, I will choose Maxwell's equations, the field equations of general relativity, Dirac's equation of relativistic quantum mechanics and a selection of mathematical theorems. It would be like saying, 'Hey, this is what we can do so far.' And ET would know where we have reached in the long quest to unravel the secrets of nature. If ever we got into a protracted dialogue and found ourselves on the same intellectual wavelength, well, then humans could follow up with cathedrals and Picassos and Beethoven symphonies, in the spirit of 'This is what *we* like. How about you?'

WHY DO SETI?

At its fiftieth anniversary, SETI remains a grand, uplifting enterprise. Its astronomers are as dedicated and positive as ever. The eerie silence has not blunted their zeal or subdued their motivation, for there is always a chance that the next observing run will finally detect something truly convincing. Meanwhile, the routine data analysis and equipment development goes on. SETI is one of very few human enterprises that really *does* take a long-term view.

In this book I have attempted to explain what we are up against when we embark on SETI, and to critically examine the hidden assumptions that underlie the present strategy. I have argued that the time has come to think much more creatively and to widen the search in novel ways, without compromising the traditional SETI programme. But even the most ardent optimist will concede that SETI is an extraordinarily long shot. All we have to go on are general scientific principles and philosophical analysis. The best that can be said is that no totally convincing argument has been given for why alien civilizations *cannot* exist.

So why do we do it? Can SETI be justified, given the poor prospects of success? I believe it can, for several reasons. First, it forces us to confront those great questions of existence that we should be thinking about anyway. What is life? What is intelligence? What is the destiny of mankind? As Frank Drake has remarked, SETI is in many ways a search for ourselves – who we are and where we fit into the universe. When we think about advanced alien civilizations, we are also glimpsing the future of mankind. The eerie silence gives us pause that such a future is by no means assured.

Fifty years is a useful benchmark, and an excellent time to evaluate the programme. It is certainly too soon to get discouraged and wind it up. As I have explained, SETI has sampled only a tiny fraction of potential habitats. But it is equally clear that the galaxy isn't obviously a hive of alien activity. 'Year after year, deep sky radio searches came up with nothing,' comments David Brin, 'none of the expected "tutorial beacons". No sign of busy interstellar communications networks. Indeed, no trace of technological civilization out there, at all.'[9] So how long should we keep at it? Because SETI's version of Moore's law implies that the search efficiency shoots up exponentially, a hundred years of silence would be very different from twice fifty years. Every additional year that produces a negative result greatly amplifies the significance of the silence, and bolsters the tentative conclusions we may draw from it.

The search for alien intelligence is an exercise in the Copernican principle which, loosely stated, says that our location in space isn't special or privileged in any way, so that what happens in our part of the universe should happen elsewhere too. The Copernican principle is not a law of nature, only a rule of thumb ('Why do we think we are so special?'). It inevitably fails at some stage, and the point at which that failure occurs is of enormous importance and interest.[10] The Copernican principle applies well to galaxies like the Milky Way, to sun-like stars within the galaxy, and – so we have recently discovered – to entire planetary systems too. What is not yet clear is whether the principle works or fails for specifically Earth-like planets across the galaxy. At the present time, scientists seem to be about equally divided between 'rare Earth' and 'common Earth' advocates, but that uncertainty may soon be rectified when the results of the Kepler planet-hunting mission become available. By contrast, we now know that within the solar system Earth is in fact rather *atypical* in its physical conditions, and that Renaissance scientists such as Huygens and Kepler were wrong to treat our sister planets as on a par with it. When it comes to biology, the case for and against the Copernican principle is finely balanced at this time. It would, however, be immediately resolved in favour of 'for' if we discover a shadow biosphere or an independent genesis of life on Mars. That doesn't take us as far as intelligence or technology though. It is possible that the Copernican principle applies all the way up to complex life, but fails when it comes to technological communities like ours. We may yet be unique.

Of course, we cannot prove a negative. We could conduct SETI for a million years without encountering any evidence of intelligent aliens, but that would not rule them out of existence. There could be all sorts of exceptional reasons why the search missed them. Nevertheless, if exhaustive searches yield nothing – if the eerie silence becomes deafening – then most people would probably think it safe to assume that we are, after all, totally alone. What then?

Concluding that we are unique in the universe would greatly amplify the value we attach to life and mind, and to the planet that sustains them. So the eerie silence could be golden. It's true that in some sense life – at least intelligent life – would have to be regarded as a freak. But does improbability diminish worth or enhance it? Certainly we should want to take better care of our planet. And we would need to take better care of ourselves too. It would be a tragedy of literally cosmic proportions if we succeeded in annihilating the one truly intelligent species in the entire universe. There is, however, a crucial caveat on which any broad conclusion about the implications for humanity hinges. In Chapter 4, I discussed whether the Great Filter lies behind us or ahead of us in time. If Earth is not just the only planet with intelligent life, but also the only planet with *any* sort of life, we will have passed through the filter already, and could be poised for a unique cosmological experiment. We might make it our mission and our destiny to spread beyond Earth, carrying the flame of life, intelligence and culture with us, to bestow this gift on countless sterile worlds. But if we discover that, although intelligence is confined to Earth, complex life is widespread, then the consequences are profoundly alarming and depressing. It implies a much higher chance that intelligence has evolved on many planets in our galaxy and others, but that it always got snuffed out, by warfare, technological accidents or any of a thousand other causes. Unless we had very good reasons for thinking we are highly atypical, then a similar fate would await us.

So the bottom line is simple. There are three possibilities, each with dramatically different implications for humanity. The first is a universe full of intelligence. That is not only exhilarating, but would promise a bright future for mankind. The second is that Earth is a unique oasis of life. That would place an awesome burden of responsibility on our shoulders, yet it would provide us with the truly cosmological mission of perpetuating a precious phenomenon – the flame of reason. But the third possibility – a universe with widespread life and nobody left bar us to celebrate it – is one that bodes very badly for our species.

MIGHT WE BE ALONE AFTER ALL? THE THREE-HATS ANSWER

People inevitably ask me, bluntly, 'Do you believe we are alone in the universe, or are there other intelligent beings out there somewhere?' In this book, I have tried to present various for-and-against arguments, but the time has come for me to get off the fence. I can do this only by wearing three hats in succession. First I shall wear my scientist hat. Do I, Paul 'The Scientist' Davies, think we are alone? As a scientist, my mind is open to new evidence and therefore not yet made up. I can assign some sort of probability for aliens to exist, based on sifting all the facts, weighted in turn by the relative importance I attach to the various arguments. When all that is put together, my answer is that we are probably the only intelligent beings in the observable universe, and I would not be very surprised if the solar system contains the only life in the observable universe. I arrive at this dismal conclusion because I see so many contingent features involved in the origin and evolution of life, and because I have yet to see a convincing theoretical argument for a universal principle of increasing organized complexity of the sort I touted in the previous chapter.

My answer may be disappointing to the reader. It is certainly disappointing to me, Paul 'The Philosopher' Davies. Wearing my second hat, and leaving science to the side, what are my feelings about the nature of a universe in which we are alone? Frankly, it makes me uneasy. I wonder what all that stuff out there is *for*, when only lowly *Homo sapiens* get to see it. Of course, my hard-headed colleagues tell me it's not *for* anything, it's just there. The idea that the universe has a purpose, they say, is just a hangover from religion.

Finally, there is Paul Davies, the human being. One of the things that influenced my choice of career was my fascination with the idea that there might be intelligent life out there somewhere. Like all teenagers, I read the flying-saucer stories, and wondered whether there might be something in them. I devoured science fiction by Arthur C. Clarke, Fred Hoyle, Isaac Asimov and John Wyndham, and pictured a galaxy pulsing with alien activity. I watched Stanley Kubrick's film *2001: A Space Odyssey* and rejoiced in the notion that humanity might have an astronomical dimension, soon to be realized. I know other scientists who followed the same path into their careers. My decades of work as a professional scientist have not diluted that wide-eyed schoolboy fascination; I would very much *like* to believe that the universe is intrinsically friendly to life and to intelligence. It suits my temperament to suppose that our humble efforts on Earth, the daily round that consumes almost all our time and energy, are part of something grander and more meaningful. I can think of no more thrilling a discovery than coming across clear evidence for extraterrestrial intelligence. In romantic moments, I like to think that all intelligent entities, biological or otherwise, enjoy a bond of fellowship that stretches across the vast reaches of space and time, and up and down the IQ ladder. Whether it is godlike quantum minds floating in the black emptiness of intergalactic space, super-cyborgs riding commandeered comets, Matrioshka brains hugging spinning black holes or humble planet-dwelling biological organisms with big brains and fancy technology, I'd like to hear from them. So wearing my 'dreamer' hat, yes, I can feel at home in a universe in which intelligent life is commonplace. This is more of a 'want' than a 'belief', but it is as far as I am prepared to go before Davies the Scientist reins me in.

And that's what makes SETI so tantalizing. We just *don't know*.

Appendix

A BRIEF HISTORY OF SETI

The year 2009 marks the 200th anniversary of the birth of Charles Darwin, and the 150th anniversary of the publication of his world-shaking book *On the Origin of Species*. It is also the fiftieth anniversary of the famous paper by Giuseppe Cocconi and Philip Morrison showing that interstellar radio communication was feasible, which paved the way for Drake's Project Ozma the following year.

For some time after Ozma, SETI was treated by the scientific community as a fringe activity. But that was set to change. In the mid-1960s, John Billingham, an ex-RAF medical doctor from the UK, began working for NASA at the Ames Laboratory in California. Through chance conversations with Ames's exobiology researchers, Billingham became enthralled with the idea of SETI. He convened an impromptu summer school, and the upshot was a detailed feasibility report called Project Cyclops, compiled by Bernard Oliver from the Hewlett-Packard Corpor -ation, and published in the early 1970s. Cyclops stimulated a flurry of activity, and observing programmes were initiated by Ohio State University, the Planetary Society, the University of California and the Jet Propulsion Laboratory in Pasadena, as well as NASA Ames and several smaller groups. The Soviet Union also had SETI projects and, to a lesser extent, so too did Western Europe and Australia. Cyclops also brought SETI to the wider public. Carl Sagan became its best-known champion. His books, articles, public lectures and highly successful television series *Cosmos* transformed the acronym SETI into a household word.

On 20 November 1984 the SETI Institute was established in Mountain View, California, close to NASA Ames, to coordinate research. (It has since moved to a location adjacent to Ames.) The US Congress finally decided in 1988 to fund a comprehensive SETI search to commemorate the 500th anniversary of Christopher Columbus's arrival in the New World. Four years later, observations began amid fanfare. Alas, this was a short-lived wonder. Within a year, Congress pulled the fiscal plug, amid a general feeling that looking for aliens was not an appropriate project for the public purse. NASA promptly stopped funding SETI. Since 1993 it has been financed almost exclusively by private donations. This enabled the SETI Institute to go ahead with Project Phoenix, a targeted search of a thousand nearby sun-like stars in both northern and southern hemispheres. Project SERENDIP at the University of California at Berkeley, and Southern SERENDIP at Parkes in Australia, also flourished. Meanwhile, public interest was elevated by the SETI@home project, in which simple software is used to adapt home computer screensavers to analyse signals from radio telescopes, holding out the faint but delicious hope that a high school student might go down in history as the person who wakes up one morning to discover ET on her PC.

Jill Tarter is currently the Director of the Center for SETI Research at the SETI Institute, and is considered by some to be the inspiration for the female lead in *Contact*. In spite of NASA's lukewarm approach to funding SETI, it actively collaborates with the SETI Institute on a wide range of research projects, including many in mainstream astrobiology. Frank Drake continues to work as an active researcher and advocate for SETI.

Bibliography

Benner, Steven, *Life, the Universe and the Scientific Method* (The Ffame Press, Gainsville, Fla., 2009)

Bennett, Jeffrey, *Beyond UFOs: The Search for Extraterrestrial Life and Its Astonishing Implications for Our Future* (Princeton University Press, Princeton, NJ, 2008)

Bracewell, Ronald, *The Galactic Club* (W. H. Freeman, San Francisco, 1975)

Chela-Flores, Julian, *A Second Genesis* (World Scientific, Singapore, 2009)

Crick, Francis, *Life Itself: Its Origin and Nature* (Touchstone, New York, 1981)

Crowe, Michael, *The Extraterrestrial Life Debate*, 1750–1900 (Cambridge University Press, Cambridge, 1986)

Davies, Paul, *The Fifth Miracle: The Search for the Origin and Meaning of Life* (Simon & Schuster, New York, 1988)

—, *The Origin of Life* (Penguin Books, London, 2003)

Dick, Steven J., *Plurality of Worlds: The Extraterrestrial Life Debate from Democritus to Kant* (Cambridge University Press, Cambridge, 1982)

— (ed.), *Many Worlds: The New Universe, Extraterrestrial Life, and the Theological Implications* (Templeton Foundation Press, West Conshohocken, Pa., 2000)

Dole, Stephen H., *Habitable Planets for Man* (Elsevier, Kidlington, 1970)

de Duve, Christian, *Vital Dust: Life as a Cosmic Imperative* (Basic Books, New York, 1995)

Dyson, Freeman, *Origins of Life* (Cambridge University Press, Cambridge, 1986)

Ekers, R. D., D. Kent Cullers and John Billingham, *SETI 2020: A Roadmap for the Search for Extraterrestrial Intelligence* (SETI Press, Mountain View, Calif., 2002)

Feinberg, Gerald and Robert Shapiro, *Life Beyond Earth: An Intelligent Earthling's Guide to Life in the Universe* (William Morrow, New York, 1980)

Gardner, James N., *Biocosm – The New Scientific Theory of Evolution: Intelligent Life is the Architect of the Universe* (Inner Ocean Publishing, Makawao, Hawaii, 2003)

Gilmour, Ian and Mark Stephton (eds.), *An Introduction to Astrobiology* (Cambridge University Press, Cambridge, 2004)

Goldsmith, Donald and Tobias Owen, *The Search for Life in the Universe*, 3rd edn (University Science Books, Sausalito, Calif., 2002)

Kurzweil, Ray, *The Age of Spiritual Machines: When Computers Exceed Human Intelligence* (Viking, New York, 1999)

Lemonick, Michael, *Other Worlds: The Search for Life in the Universe* (Simon & Schuster, New York, 1998)

McConnell, Brian S., *Beyond Contact: A Guide to SETI and Communicating with Alien Civilizations* (O'Reilly Media, Inc., Sebastopol, Calif., 2001)

Morris, Simon Conway, *Life's Solution: Inevitable Humans in a Lonely Universe* (University of Cambridge, Cambridge, 2003)

Plaxco, Kevin W. and Michael Gross, *Astrobiology: A Brief Introduction* (The Johns Hopkins University Press, Baltimore, 2006)

Sagan, Carl, *Contact* (Simon & Schuster, New York, 1985; Century Hutchinson, London, 1985)

—, *Cosmos* (Random House, New York, 1980; Macdonald & Co., London, 1981)

Shapiro, Robert, *Origins: A Skeptic's Guide to the Creation of Life on Earth* (Summit Books, New York, 1986)

Shermer, Michael, *Why People Believe Weird Things: Pseudoscience, Superstition, and Other Confusions of Our Time* (W. H. Freeman, San Francisco, 1997)

Shostak, Seth, *Sharing the Universe: Perspectives on Extraterrestrial Life* (Berkeley Hills Books, Albany, Calif., 1998)

—, *Confessions of an Alien Hunter: A Scientist's Search for Extraterrestrial Intelligence* (National Geographic, Washington, DC, 2009)

Shuch, H. Paul, *Tune into the Universe: A Radio Amateur's Guide to the Search for Extraterrestrial Intelligence* (American Radio Relay League, Hartford, Conn., 2001)

Ward, Peter and Donald Brownlee, *Rare Earth: Why Complex Life is Uncommon in the Universe* (Copernicus, New York, 2000)

Webb, Stephen, *If the Universe is Teeming with Aliens . . . Where is Everybody? Fifty Solutions to Fermi's Paradox and the Problem of Extraterrestrial Life* (Copernicus, New York, 2002)

Notes

Preface

1. Today, the significance of Jansky's discovery is recognized by the name assigned to the unit of radio flux – the jansky.
2. 'Searching for interstellar communications', by Giuseppe Cocconi and Philip Morrison, *Nature*, vol. 184 (1959), p. 844.

1. Is Anybody Out There?

1. The unit MHz stands for 'megahertz', hertz being a measure of frequency named after the German physicist Heinrich Hertz. It is equivalent to 1 cycle per second. 1 MHz is 1 million hertz. 1 gigahertz, written GHz, is 1 billion hertz, or 1,000 MHz. The frequency 1,420 MHz corresponds to a wavelength of 21 cm. An automatic device enabled Drake to scan a narrow-frequency range around 1,420 MHz.
2. A more realistic description of how SETI flaps work in practice is given by Seth Shostak in his book *Confessions of an Alien Hunter: A Scientist's Search for Extraterrestrial Intelligence* (National Geographic, 2009).
3. Motion of the source or receiver shifts the frequency in a time-varying manner because of the Doppler effect. Without correction, an alien radio signal would drift out of a fine-tuned frequency band in just a few minutes.
4. H. G. Wells, *The War of the Worlds* (Heinemann, London, 1898), p. 4.
5. For an endorsement of the information motive, see, for example, T. B. H. Kuiper and M. Morris, 'Searching for extraterrestrial civilizations', *Science,* vol. 196 (1977), p. 616; D. G. Stephenson, 'Models of interstellar exploration,' *Quarterly Journal of the Royal Astronomical Society*, vol. 23 (1982), p. 236.

6. Foreword by Frank Drake in *Confessions of an Alien Hunter: A Scientist's Search for Extraterrestrial Intelligence* by Seth Shostak (National Geographic, 2009), p. ix.
7. Carl Sagan, *Cosmos* (Random House, New York, 2002), p. 339.
8. http://www.meteorlab.com/METEORLAB2001dev/metics.htm#Thomas
9. A good example from particle physics was the discovery of the *W* and *Z* particles at CERN in the early 1980s. The discoveries were announced after only a handful of actual 'events' had been detected in the Large Electron Positron collider. Few physicists quibbled, because an excellent theory predicting *W* and *Z* had been worked out a decade earlier, and gave very specific quantitative predictions of what the new particles would be like.
10. Rupert Sheldrake has come closest to producing a scientific theory of something like telepathy, one that makes broad falsifiable predictions, but it still lacks a credible physical basis and a proper mathematical model of the mechanism involved. For a review, see Rupert Sheldrake, *The Sense of Being Stared At: And Other Aspects of the Extended Mind* (Crown, New York, 2003).
11. In mathspeak, the prior probability of a communicating civilization in our galaxy is likely to be 'bimodal' – either very close to zero or very close to 1 (a probability of 1 is a certainty). Note that it is not then legitimate to assign a prior probability of ½ (being the average of 0 and 1) in the absence of any other evidence, any more than we can say there is a 50–50 chance of there being an afterlife on the basis that about half the population think there is and the other half think there isn't.
12. Ezekiel 1:4–28.
13. Democritus according to Hippolytus, *Refutation of the Heresies* I 13 2, in Hermann Diels and Walther Kranz, *Die Fragmente der Vorsokratiker* (Weidmann, Zurich, 1985), vol. 2, section 68 A 40, p. 94. Translation from W. K. C. Guthrie, *A History of Greek Philosophy: Presocratic Tradition from Parmenides to Democritus* (Cambridge University Press, Cambridge, 1965), vol. 2, p. 405.
14. *The Roman Poet of Science, Lucretius: De Rerum Natura* Book II (trans. Alban Dewes Winspear, The Harbor Press, New York, 1955).
15. *Kepler's Conversation with Galileo's Sidereal Messenger* (trans. Edward Rosen, Johnson reprint, New York and London, 1965), p. 42.
16. http://ufos.nationalarchives.gov.uk/
17. Edward Condon, *Scientific Study of Unidentified Flying Objects* (University of Colorado, Boulder, 1968).
18. J. B. S. Haldane, *Possible Worlds: And Other Essays* (Chatto and Windus, London, 1932), p. 286.

2. Life: Freak Side-Show or Cosmic Imperative?

1. *Washington Post*, 20 July 2008.
2. Francis Crick, *Life Itself: Its Origin and Nature* (Simon & Schuster, New York, 1981), p. 88.
3. Jacques Monod (trans. A. Wainhouse), *Chance and Necessity* (Collins, London, 1972), p. 167.
4. George Gaylord Simpson, 'The non-prevalence of humanoids', *Science*, vol. 143 (1964), p. 769.
5. Christian de Duve, *Vital Dust: Life as a Cosmic Imperative* (Basic Books, New York, 1995).
6. http://www.telegraph.co.uk/scienceandtechnology/science/space/4629672/AAAS-One-hundred-billion-trillion-planets-where-alien-life-could-flourish.html
7. J. William Schopf and Bonnie M. Packer, 'Newly discovered early Archean (3.4–3.5 Ga Old) microorganisms from the Warrawoona Group of Western Australia', *Origin of Life and Evolution of Biospheres*, vol. 16, nos. 3–4 (1986), p. 339.
8. A. Allwood, 'Stromatolite reef from the Early Archaean Era of Australia', *Nature*, 8 June 2006, p. 714.
9. I discussed this process in detail in my book *The Fifth Miracle* (Simon & Schuster, New York, 1998; Allen Lane, The Penguin Press, London, 1998), published in a revised edition in the UK under the title *The Origin of Life* (Penguin, London, 2003).
10. Gerda Horneck, et al., 'Microbial rock inhabitants survive hypervelocity impacts on Mars-like host planets: first phase of lithopanspermia experimentally tested', *Astrobiology*, vol. 8, no. 1 (2008), p. 17.
11. Fred Hoyle, *The Intelligent Universe* (Michael Joseph, London, 1983), pp. 18–19.
12. George Whitesides, 'The improbability of life', in John D. Barrow, Simon Conway Morris, Stephen J. Freeland and Charles L. Harper (eds.), *Fitness of the Cosmos for Life: Biochemistry and Fine-Tuning* (Cambridge University Press, Cambridge, 2004), p. xiii.
13. Ibid., p. xv.
14. Ibid., p. xvii.
15. Ibid.
16. There may be other combinations of molecules, also random in the sense of being pattern-less, that would represent a different form of life. The point is that biologically functional molecular sequences occupy a tiny overall

fraction of the total sequence space, even if there are very many disconnected regions representing possible biological functionality.

17. And just to be completely clear, when I use the colloquial term 'near-miracle' I am *not* suggesting that the origin of life was due to some sort of divine intervention. I think it was a perfectly natural process, though perhaps an exceedingly improbable one.
18. Let me be explicit: if you examine a string of fifty amino acids and try to guess on mathematical grounds alone from the prior sequence what the next amino acid will be, then you will be right only to the extent of pure chance. The same goes for base-pair sequences in DNA.
19. Paul Davies, *The Cosmic Blueprint*, rev. edn (Templeton Foundation Press, West Conshohocken, Pa., 2004). See also the final chapter of *The Fifth Miracle*.
20. A good introduction to this field is William Poundstone, *The Recursive Universe* (William Morrow, New York, 1996). A more in-depth (and contentious) discussion may be found in Stephen Wolfram, *A New Kind of Science* (Wolfram Media, Champaign, Ill., 2002).
21. A. G. Cairns-Smith, *Seven Clues to the Origin of Life* (Cambridge University Press, Cambridge, 1986).
22. I discuss a specific model in 'It's a quantum life', *Physics World*, vol. 22, no. 7 (2009), p. 24.
23. Mars remains the favourite, but Europa, a moon of Jupiter, is another possible abode for primitive life. It is an ice-covered body, with an ocean of liquid water beneath, warmed by a process called tidal flexing. As it orbits Jupiter, Europa gets deformed by the giant planet's gravitational field, elongating its entire body, including the solid core. That generates a lot of frictional heat. Another body of great interest is Titan, a large moon of Saturn. In 2008 a small probe called Huygens was parachuted to Titan's surface, and revealed a frigid world with rivers and lakes of liquid methane and ethane, rocks of water ice, and a thick atmosphere of petrochemical smog. This lethal cocktail would finish off terrestrial organisms in no time at all, but some scientists have conceived of exotic low-temperature life for which liquid water is replaced by a different solvent, and metabolism hinges on the conversion of acetylene to methane.
24. Unless, by a perverse stroke of bad luck, Mars hosts two forms of life, with opposite chirality and equal population density.

3. A Shadow Biosphere

1. Kevin Maher and David Stevenson, 'Impact frustration of the origin of life', *Nature*, vol. 331 (1988), p. 612.
2. I mooted this idea in 1988 in my book *The Fifth Miracle*. A detailed study is reported in Lloyd E. Wells, John C. Armstrong and Guillermo Gonzalez, 'Reseeding of early earth by impacts of returning ejecta during the late heavy bombardment', *Icarus*, vol. 162, no. 1 (2003), p. 38.
3. The term 'shadow biosphere' was coined by Carol Cleland and Shelley Copley of the University of Colorado in their paper 'The possibility of alternative microbial life on Earth', *International Journal of Astrobiology*, vol. 4 (2005), p. 165.
4. Richard Dawkins, *The Ancestor's Tale* (Houghton Mifflin, Boston, 2004; Weidenfeld & Nicolson, London, 2004).
5. Paul C. W. Davies and Charley H. Lineweaver, 'Search for a second sample of life on Earth', *Astrobiology*, vol. 5, no. 2 (2005), p. 154.
6. Paul Davies, Steven Benner, Carol Cleland, Charley Lineweaver, Chris McKay and Felisa Wolfe-Simon, 'Signatures of a shadow biosphere', *Astrobiology*, vol. 9, no. 2 (2009), p. 1.
7. Stephen Jay Gould, 'Planet of the Bacteria', *Washington Post Horizon*, vol. 119 (1996), p. 344.
8. This is something of a simplification. Whilst some organisms can use only the inorganic gases hydrogen and carbon dioxide as input, others make indirect use of surface biology through dissolved oxygen or organic substances that sink down from sunlit layers near the sea surface.
9. Thomas Gold, *The Deep Hot Biosphere* (Springer, New York, 1999). For an up-to-date review, see Bo Barker Jorgensen and Steven D'Hondt, 'A starving majority deep beneath the sea floor', *Science*, vol. 314 (2006), p. 932.
10. For a review, see my book *The Fifth Miracle*.
11. T. O. Stevens and J. P. McKinley, 'Lithoautotrophic microbial ecosystems in deep basalt aquifers', *Science*, vol. 270 (1995), p. 450; D. R. Lovley, 'A hydrogen-based subsurface microbial community dominated by methanogens', *Nature*, vol. 415 (2002), p. 312; L. H. Lin, et al., 'Long-term sustainability of a high-energy, low-diversity crustal biome', *Science*, vol. 314 (2006), p. 479.
12. Astrobiologists speculate that there may be similar subsurface ecosystems on Mars – hence the flurry of excitement when methane was discovered in the Martian atmosphere a few years ago.
13. By some definitions, viruses themselves are alive, so a weird virus would

alone count as a discovery of weird life. Viruses are a marginal case, because they cannot reproduce without the help of a cell, so they are not autonomous organisms. But if we find weird viruses, then weird cells are unlikely to be far away.

14. If Gil's revamped Labelled Release experiment works well on Earth, the next step would be to send it to Mars to clear up the Viking mystery once and for all.
15. As I already explained, when I call these interlopers 'aliens', it is in the sense of being 'other'. It does not imply they 'came from outer space' to use sci-fi jargon, although they may have done. They may have come from Mars; but so might our own distant ancestors.
16. P. C. W. Davies, E. V. Pikuta, R. B. Hoover, B. Klyce and P. A. Davies, 'Bacterial utilization of L-Sugars and D-amino acids,' proceedings of SPIE's 47th annual meeting, San Diego, August 2006, 63090A.
17. Steven Benner, *Life, the Universe and the Scientific Method* (The Ffame Press, Gainsville, Fla., 2009).
18. Ariel Anbar, Paul Davies and Felisa Wolfe-Simon, 'Did nature also choose arsenic?', *International Journal of Astrobiology*, Vol. 8 (2009), p. 69.
19. In technical language, it offers a redox potential by permitting arsenate to be reduced to arsenite, releasing energy as a result.
20. For example, mass spectrometry, which can measure the relative weights of molecules and thereby sort organics into categories.
21. There is a further complicating factor. In discussing 'the origin of life' I have tacitly assumed that there is a clear demarcation between the 'non-living' and 'living' states, so that biogensis is a well-defined event. But this may be an unwarranted simplification. There may be no clear line separating life from non-life, merely a seamless and extended chemical pathway to states of greater and greater complexity.
22. I am grateful to Chris McKay and Felisa Wolfe-Simon for drawing my attention to those examples.
23. Brent C. Christner, Cindy E. Morris, Christine M. Foreman, Rongman Cai and David C. Sands, 'Ubiquity of biological ice nucleators in snowfall', *Science*, vol. 319 (2008), p. 1214.
24. R. L. Folk, 'SEM imaging of bacteria and nanobacteria in carbonate sediments and rocks', *Journal of Sedimentary Petrology*, vol. 63 (1993), p. 990.
25. Philippa J. R. Uwins, Richard I. Webb and Anthony P. Taylor, 'Novel nano-organisms from Australian sandstones', *American Mineralogist*, vol. 83 (1998), p. 1541.
26. E. O. Kajander and N. Ciftcioglu, 'Nanobacteria: an alternative mechanism for pathogenic intra- and extracellular calcification and stone formation',

Proceedings of the National Academy of Sciences, vol. 95 (1998), p. 8274.

27. Benner, *Life, the Universe and the Scientific Method*, pp. 122–3.
28. For a detailed account of the Mars meteorite, see my book *The Fifth Miracle*.
29. J. Martel and J. D.-E. Young, 'Purported nanobacteria in human blood as calcium carbonate nanoparticles', *Proceedings of the National Academy of Sciences*, 8 April 2008, vol. 105, no. 14 (2008), p. 5549.
30. Jocelyn Selim, 'Venter's ocean genome voyage', *Discover* online, 27 June 2004.

4. How Much Intelligence is Out There?

1. Charles Darwin, *On the Origin of Species* (John Murray, London, 1859), final page.
2. H. J. Jerison, *Evolution of the Brain and Intelligence* (Academic Press, New York, 1973). The *expected* brain to body size ratio is computed using a scaling law averaging over many animals that assumes the brain mass should vary like the 2/3 power of the body mass, that being the surface area to volume ratio. This assumption, and indeed the very notion of EQ as a useful measure of intelligence, has been criticized. See, for example, Robert O. Deaner, Karin Isler, Judith Burkart and Carel van Schaik, 'Overall brain size, and not encephalization quotient, best predicts cognitive ability across non-human primates', *Brain, Behavior and Evolution*, vol. 70 (2007), p. 115.
3. See, for example, http://serendip.brynmawr.edu/bb/kinser/Int3.html.
4. This is the type of growth, characteristic of all unrestrained expansion, where a quantity doubles in a fixed time. See, for example, D. A. Russell, 'Exponential evolution: implications for intelligent extraterrestrial life', *Advances in Space Research*, vol. 3, (1983) p. 95.
5. Stephen Jay Gould, *Wonderful Life* (Norton, New York, 1990).
6. See, for example, Simon Conway Morris, *Life's Solution: Inevitable Humans in a Lonely Universe* (Cambridge University Press, Cambridge, 2003). Another factor that weakens Gould's argument is its neglect of feedback mechanisms that serve to reinforce evolutionary trends. See Robert Wright, *Nonzero: The Logic of Human Destiny* (Pantheon, New York, 2000).
7. Lineweaver articulates his argument in a review of Peter Ulmschneider's book *Intelligent Life in the Universe*, in *Astrobiology*, vol. 5, no. 5 (2005), p. 658. See also C. H. Lineweaver, 'Paleontological tests: human-like intelligence is not a convergent feature of evolution', in J. Seckbach and M. Walsh (eds.), *From Fossils to Astrobiology* (Springer, New York, 2009), p. 353.
8. Christopher P. McKay, 'Time for intelligence on other planets', in Laurance

R. Doyle (ed.), *Circumstellar Habitable Zones, Proceedings of the First International Conference* (Travis House Publications, Menlo Park, Calif., 1996), p. 405.

9. See, for example, Lori Marino, 'Convergence of complex cognitive abilities in Cetaceans and Primates', *Brain, Behavior and Evolution*, vol. 59 (2002), p. 21.

10. See, for example, Mircea Eliade (trans. Willard R. Trask), *The Myth of the Eternal Return* (Princeton University Press, Princeton, NJ, 1971).

11. Joseph Needham and collaborators, *Science and Civilization in China*, 7 vols. (Cambridge University Press, Cambridge, 1954–).

12. The relevant number for SETI is actually the rate of star formation some billions of years ago.

13. On the other hand, rogue planets may not offer good prospects for advanced life, although we cannot be sure. The Drake equation also omits the possibility that some planets may acquire life and/or intelligence on account of being colonized rather than it arising *de novo*. This is a topic I discuss in Chapter 6.

14. I am ignoring the light travel time when I say 'now', as the basic argument is unaffected.

15. Michael Shermer, 'Why ET hasn't called', *Scientific American*, 15 July 2002.

16. A good example of what is presumably a coincidence of two causally independent timescales is the lunar cycle and the human menstrual cycle, both about twenty-eight days.

17. Carl Sagan, 'The abundance of life-bearing planets', *Bioastronomy News*, vol. 7, no. 4 (1995), p. 1.

18. Brandon Carter, 'The anthropic principle and its implications for biological evolution', *Philosophical Transactions of the Royal Society of London*, vol. A 310 (1983), p. 347.

19. Robin Hanson, 'The great filter: are we almost past it?', http://hanson.gmu.edu/greatfilter.html (1998).

20. As I explained earlier, this hypothesis was widely accepted when Carter formulated his argument in about 1980.

21. Brandon Carter, 'Five or six step scenario for evolution?', *International Journal of Astrobiology*, vol. 7 (2008), p. 177.

22. In the absence of any special reason to the contrary, we should assume that humans are typical observers. Carter's argument is consistent with that typicality assumption, for suppose we envisage an enormous volume of space – much bigger than the observable universe – and focus on the sub-class of all (according to Carter, exceedingly rare) planets with intelligent observers. Then Earth should be a typical member of that sub-class; and as

far as we know that is the case. By contrast, if Carter is wrong and intelligent life is very likely and quick to arise, then because humans were so tardy in evolving on Earth, we would be *atypical* observers.

23. An alternative explanation, of course, is that we are not alone, but the aliens have not so far manifested their existence in a way that has been noticed by us. They may have ceased radio emissions after a brief duration, for example.
24. See, for example, John Leslie, *The End of the World: The Science and Ethics of Human Extinction* (Routledge, London, 1996), and Martin Rees, *Our Final Century* (Arrow Books, London, 2004).
25. Nick Bostrom, 'Where are they? Why I hope the search for extraterrestrial life will find nothing', *MIT Technology Review*, May/June issue (2008), pp. 72, 77.

5. New SETI: Widening the Search

1. Abraham Loeb and Matias Zaldarriaga, 'Eavesdropping on radio broadcasts from galactic civilizations with upcoming observatories for redshifted 21cm radiation', astro-ph/0610377 (October 2006). The authors estimate that much more powerful military-radar-strength pulses might be detectable with the SKA from as far away as 650 light years, for a one-month integration time.
2. The sensitivity of an instrument depends not only on the collecting area, but also on the computer algorithm used to extract the signal from the noise. Recent work by Claudio Maccone suggests that a technique known as the KL transform, named after the mathematicians Kari Karhunen and Maurice Loève who proposed it in 1949, may lead to an improvement in sensitivity by a factor of up to a thousand.
3. John G. Learned, Sandip Pakvasa and A. Zee, 'Galactic neutrino communication', *Physics Letters B*, vol. 671, no. 1 (2009), p. 15.
4. Modern lighthouse signals are encoded with identifying information too.
5. The examples at the start of this section fall under the category of 'active SETI' or METI – messaging extraterrestrial intelligence – a contentious subject I shall return to in Chapter 9.
6. Regarding my earlier remarks about energy conservation being an anthropocentric concern, I distinguish between energy as not being a priority issue for aliens and their deliberately squandering it for no good purpose. Even if energy is cheap, you still have to acquire it.
7. Gregory Benford, James Benford and Dominic Benford, 'Cost optimized interstellar beacons: SETI', to be published.

8. In 1989, Sagan and Horowitz analysed thirty-seven unexplained pulses, and although the sources showed a tendency to cluster in the galactic plane, the authors concluded they were not strong evidence of ETI.
9. M. J. Rees, 'A better way of searching for black-hole explosions?', *Nature*, vol. 266 (1977), p. 333.
10. The inner core of the galaxy, within about 1,000 light years of the centre, is an unpromising location for advanced life, for reasons I shall explain in the next section.
11. Robert A. Rohde and Richard A. Muller, 'Cycles in fossil diversity', *Nature*, vol. 434 (2005), p. 208.
12. Mikhail V. Medvedev and Adrian L. Melott, 'Do extragalactic cosmic rays induce cycles in fossil diversity?', *Astrophysical Journal*, vol. 664 (2007), p. 879.
13. Anxious readers should rest assured that the solar system is currently close to the galactic plane and well away from the danger zone.
14. A neutron star is the remnant of the core of a large star that has imploded under its own immense weight to form an exceedingly dense ball of neutrons, typically only a few kilometres across, but with a mass exceeding that of the sun.
15. William H. Edmondson and Ian R. Stevens, 'The utilization of pulsars as SETI beacons', *International Journal of Astrobiology*, vol. 2, no. 4 (2003), p. 231.
16. I include computer intelligence in the definition of alien intelligence, for reasons I shall discuss further in Chapter 8. The conversation would be directly with the probe and not with the probe's dispatchers.
17. Ronald N. Bracewell, 'Communications from superior galactic communities', *Nature*, vol. 186 (1960), p. 670. Reprinted in A. G. Cameron (ed.), *Interstellar Communication* (W. A. Benjamin, Inc., New York, 1963), p. 243.
18. This is an orbit with a period of one day, so that the satellite appears to remain stationary above a fixed point on Earth. Television satellites do this.
19. There are also Earth–Moon Lagrange points, which have been the subject of limited searches.
20. For many decades 5- to 10-second radio-broadcast echoes have been detected, and remain something of a mystery. See Volker Grassmann, 'Long-delayed radio echoes: observations and interpretations', *VHF Communications*, vol. 2, 109 (1993).
21. John von Neumann, edited and completed by Arthur W Burks, 'The theory of self-reproducing automata', (University of Illinois Press, Urbana, Ill., 1966).
22. The text of his address, which was delivered at the California Institute of

Technology, is reproduced at http://www.mrs.org/s_mrs/doc.asp?CID=8969&DID=195829.

23. This scenario, and the term 'gray goo', was introduced by the nanotechnology pioneer Eric Drexler in his 1986 book *Engines of Creation* (Doubleday, New York, 1986; Anchor Books, Peterborough, 1986).

24. Strictly speaking a virus is not a von Neumann machine because it cannot reproduce unaided; it must infect a host cell to manufacture replicas.

25. This basic idea was discussed many years ago by Francis Crick, although his speculation was that aliens had propelled microbes across space together with a 'starter kit' to incubate them, with the purpose of seeding Earth and other planets with life, rather than conveying a message. See Francis Crick and Leslie E. Orgel, 'Directed panspermia', *Icarus*, vol. 19, 341 (1973), and Francis Crick, *Life Itself: Its Origin and Nature* (Simon & Schuster, New York, 1981).

26. Another strategy would be to insert 'DNA-friendly' informational molecules that would not themselves be DNA; rather, they would be made up of molecular building blocks other than the standard A,G,C,T toolkit of known life, and chosen for their chemical stability and lower mutation rate. For this idea to work, sequences of these building blocks would still have to be accurately replicated by the biochemical machinery of standard life.

27. This idea has been investigated over many years by Fred Hoyle and Chandra Wichramasinghe. See, for example, F. Hoyle and N. C. Wickramasinghe, *Astronomical Origins of Life*, in *Astrophysics and Space Science*, vol. 268 (2000), which reprints much of their earlier work.

28. H. Yokoo and T. Oshima, 'Is bacteriophage phi X174 DNA a message from an extraterrestrial intelligence?', *Icarus*, vol. 38 (1979), p. 148.

6. Evidence for a Galactic Diaspora

1. From Arthur Conan Doyle, *The Sign of the Four*, in *Lippincott's Monthly Magazine* (February 1890).

2. Stephen Webb, *If the Universe is Teeming with Aliens . . . Where Is Everybody? Fifty Solutions to Fermi's Paradox and the Problem of Extraterrestrial Life* (Copernicus Books, New York, 2002).

3. Ronald Bracewell, *The Galactic Club* (Freeman, San Francisco, 1975).

4. Stephen Hawking, 'Chronology protection conjecture', *Physical Review D*, vol. 46 (1992), p. 603.

5. No evidence and precious little theoretical support, either, for astronaut-sized wormholes. Ultra-microscopic ones are more feasible.

6. Some people pin their hopes on space privateers. So far the private sector space programme is limited to joyrides, but in the event of the full commercialization of space, private industry could overtake government agencies in space exploration/tourism.
7. George Dyson, *Project Orion: The True Story of the Atomic Spaceship* (Henry Holt, New York, 2002).
8. Seth Shostak, *Confessions of an Alien Hunter: A Scientist's Search for Extraterrestrial Intelligence* (National Geographic, Washington, DC, 2009), p. 264.
9. Geoffrey Landis, 'The Fermi paradox: an approach based on percolation theory', *Journal of the British Interplanetary Society*, vol. 51 (1998), p. 163.
10. Robin Hanson, 'The rapacious hardscrapple frontier', in Damien Broderick (ed.), *Year Million: Science at the Far Edge of Knowledge* (Atlas Books, Ashland, Ohio, 2008), p. 168.
11. This consideration is irrelevant, however, if the colonists were non-biological machines. In that case, Earth's indigenous biology might prove attractive as raw material for making bio-machines to assist the colonists' enterprises. It is fascinating to speculate whether the descendants of these discarded alien creations are still around, forming a shadow biosphere awaiting detection. But there is clearly an even more dramatic possibility, which is that the aliens visited Earth 3.5 billion years ago and created terrestrial life *ab initio*, in the form of clever nanomachines to help with the chores. If they released these synthetic organisms into the environment and didn't clean up properly, it would have a bizarre implication: we could be the distant descendants of alien bio-trash left behind when the expedition moved on!
12. More plausibly, the probability will rise slowly over time as the number of habitable planets accumulates, so the chance of alien visitation should be weighted somewhat in favour of more recent epochs, but not enough to contradict the broad conclusion I have drawn.
13. An analogous suggestion on these lines was made by Frank Drake, who pointed out that an alien civilization might create a beacon by dumping a large quantity of a rare element with a short half-life into its parent star. A good candidate is technetium, which does not occur naturally on Earth (although it can be manufactured). The presence of technetium lines in the spectrum of a star would strongly suggest the presence of a technological civilization.
14. Alan Weisman, *The World Without Us* (Picador, London, 2007).
15. Trace amounts of Pu^{244} isotope have been found on the Moon, and at Oklo, but nothing concentrated enough to raise eyebrows. A certain amount was present when the solar system formed, but most of it has now decayed.
16. Greg Bear, *The Forge of God* (Tor Books, New York, 2001).

17. Olaf Stapledon, *Star Maker* (Methuen, London, 1937).
18. I well remember a sober lunch conversation in 1975 in the student cafeteria of the London School of Economics, near King's College, where I was at the time working in the Mathematics Department. My colleague Chris Isham reported on a claim that a balloon-borne cosmic ray experiment had detected a magnetic monopole, and we gloomily reflected on the potential of these particles for weapons of mass destruction.
19. For a popular account, see Dennis Overbye, 'A whisper, perhaps, from the universe's dark side', *The New York Times*, 25 November 2008.
20. Curiously, cosmic strings have been invoked as a possible explanation for Lorimer's pulse (see p. 100), although no suggestion has been made that it involved alien technology.

7. Alien Magic

1. Freeman Dyson, 'Search for artificial stellar sources of infrared radiation', *Science*, vol. 131 (1960), p. 1667.
2. Richard A. Carrigan Jr, 'IRAS-based whole-sky upper limit on Dyson spheres', in astro-ph 0811.2376.
3. For a discussion, see Richard Dawkins, *The Blind Watchmaker* (Norton, New York, 1986).
4. David Bohm, *Wholeness and the Implicate Order* (Routledge, London, 1996).
5. Lawrence Krauss, *The Physics of Star Trek* (Harper & Row, New York, 1996).
6. See my book *How to Build a Time Machine* (Penguin/Viking, London and New York, 2002) for a review of the problems.
7. Microscopic short-lived wormholes might just be possible, and could conceivably be made in particle accelerators like the Large Hadron Collider at CERN.
8. Arthur Eddington, *The Nature of the Physical World* (Cambridge University Press, Cambridge, 1928), p. 74.
9. For a discussion of the way in which the expansion of the universe is speeding up, see my book *The Goldilocks Enigma* (Penguin, London, 2006, and Houghton Mifflin, Boston, 2008).
10. Quantum mechanics predicts a finite probability for the universe to tunnel from one vacuum state to a lower one. If this happened at a given point in space, it would create a bubble that would expand out at nearly the speed of light, engulfing and obliterating all matter in its path. A nice science fiction

story along these lines is Stephen Baxter's *Manifold: Time* (Del Ray, New York, 2000).

11. Negative energy and pressure is related to the exotic matter needed to stabilize wormholes.

8. Post-Biological Intelligence

1. S. Butler in *Canterbury Press*, 13 June 1863.
2. *The Times online*, 24 April 2007.
3. If they are, it is far from obvious that humans would choose genetic enhancement of the Mekon-resembling variety. I can well imagine the clamour for glamour would take precedence. Or perhaps sporting prowess.
4. It is also easy to imagine a nightmare society of monsters and suffering.
5. Alan Turing, 'Can machines think?', *Mind*, vol. 59 (1950), p. 433.
6. I am side-stepping the depressing prospect that humans may try to program the machines to fight their own literal and metaphorical battles, even when the machines outsmart them.
7. I am not alone in advocating a post-biological universe dominated by 'machine' intelligence. The historian of science Steven Dick has developed the idea in detail. See his essay 'Cultural evolution, the post-biological universe and SETI', *International Journal of Astrobiology*, vol. 2, no. 1 (2003), p. 65.
8. An ATS differs from the Blue Brain simulation I discussed earlier, which would have a personal identity. The latter is a simulation of a real biological brain, not a post-biological entity.
9. http://www.aeiveos.com:8080/~bradbury/MatrioshkaBrains/MatrioshkaBrainsPaper.html
10. Dominated in intellectual terms, that is. In terms of sheer numbers, smaller brains/computers will proliferate much faster.
11. In fact, a superposition is more general than I have described, because the admixture of heads and tails can be a complex number.
12. The results of a quantum computation evade the generic vagaries of quantum uncertainty only if certain specially selected states are used at the point of input and output. A handful of quantum algorithms have been discovered for solving special classes of mathematical problems making use of this.
13. For an introduction, see *The Feynman Processor* by Gerard Milburn (Basic Books, New York, 1999).

9. First Contact

1. Stephen Baxter, 'Renaissance v. revelation: the timescale of ETI signal interpretation'. *Journal of the British Interplanetary Society*, vol. 62 (2009), p. 131.
2. http://www.coSETI.org/SETIprot.htm
3. A graphic account of these events is given by Seth Shostak, who was there at the time, in his book *Confessions of an Alien Hunter: A Scientist's Search for Extraterrestrial Intelligence* (National Geographic, Washington, DC, 2009).
4. S. Shostak and C. Oliver, 'Immediate reaction plan: a strategy for dealing with a SETI detection', in G. Lemarchand and K. Meech (eds.), *Bioastronomy 99: A New Era in the Search for Life*, ASP Conference Series, vol. 213 (2000), p. 635.
5. Ibid., p. 636.
6. Ibid., p. 635.
7. For a vivid and critical account, see Frank Close, *Too Hot to Handle: The Story of the Race for Cold Fusion* (W. H. Allen, London, 1990).
8. Which very nearly happened on 13 January 2004, when astronomers in the US computed a one-in-four chance that a 500-metre-wide asteroid might hit the Earth within thirty-six hours. They sensibly held off calling the White House in the middle of the night until improved data showed all was well.
9. http://impact.arc.nasa.gov/news_detail.cfm?ID=122
10. *Acta Astronautica*, vol. 21 (1990), no. 2, p. 153.
11. A famous hoax, known as the EQ Peg affair, occurred on 28 October 1998, when an anonymous amateur astronomer in Britain claimed to have picked up a signal from the relatively nearby star EQ Pegasi using a small radio dish belonging to his employer, a UK electronics company. None of the established SETI protocol was observed. The BBC broke the story, which then attracted major media attention around the world. Professional SETI scientists were suspicious from the start. Unable to verify the signal, Paul Shuch and his SETI League colleagues discovered that the signal images were fabricated using commercially available software. When the SETI League and SETI Institute debunked the claim, the tabloids predictably accused them of a sinister cover-up. At no stage did any government agency show the slightest interest.
12. The iconic picture of earthrise from the Moon, taken by the Apollo astronauts, boosted the rise of environmentalism in the 1970s by dramatically emphasizing how precious and how isolated is our little haven of life in a hostile and often violent universe.

13. Carl Sagan, *The Cosmic Connection* (Hodder and Stoughton, London, 1974), pp. 218–19.
14. P. W. Atkins, *The Second Law*, 2nd edn (Scientific American Books, New York, 1994), p. 200.
15. I have discussed these ideas in greater depth in my book *The Cosmic Blueprint* (Simon & Schuster, New York, 1988). See also Stuart Kauffman, *At Home in the Universe: The Search for the Laws of Self-Organization and Complexity* (Oxford University Press, Oxford, 1996).
16. Bertrand Russell, *Mysticism and Logic* (Barnes & Noble, New York, 1917), pp. 47, 48.
17. An in-depth discussion of the philosophy of progress can be found in John Barrow and Frank Tipler, *The Anthropic Cosmological Principle* (Oxford University Press, Oxford, 1986).
18. Martin Rees, *Our Final Hour* (Basic Books, New York, 2003); *Our Final Century: Will the Human Race Survive the Twenty-First Century?* (William Heinemann, London, 2003).
19. See, for example, Freeman Dyson, 'Our biotech future', *The New York Review of Books*, vol. 51, no. 12 (19 July 2007).
20. Ray Kurzweil, *The Singularity is Near* (Viking, New York, 2005).
21. In this section I shall bypass the possibility that ET is some sort of machine intelligence or even an ATS, as it is hard enough to discuss the moral dimension of alien biological organisms.
22. There is the third solution, which is that the aliens are saved by some other mode of divine intervention about which we cannot guess. This response, however, simply puts the problem in the 'too hard' basket.
23. http://padrefunes.blogspot.com/2008/05/extraterrestrial-is-my-brother.html
24. Ted Peters and Julie Froehlig, 'The Peters ETI religious crisis survey', 2008, http://www.counterbalance.net/etsurv/index-frame.html.
25. That at least is the folklore. Ernan McMullin, a philosopher of religion, has criticized it as simplistic.
26. http://www.daviddarling.info/encyclopedia/W/Whewell.html
27. William Whewell, *The Plurality of Worlds* (Gould and Lincoln, Boston, 1854).
28. Emanuel Swedenborg, *Earths in the Universe* (The Swedenborg Society, London, 1970).
29. Ibid., p. 47.
30. Ibid., p. 60.
31. Ibid., p. 3.
32. E. A. Milne, *Modern Cosmology and the Christian Idea of God* (Clarendon Press, Oxford, 1952), p. 153.

33. Milne's proposal was slammed in 1956 by E. L. Mascall, a philosopher and priest, in favour of multiple incarnations to save any 'rational corporeal beings who have sinned and are in need of redemption'. See E. L. Mascall, *Christian Theology and Natural Science* (Ronald Press, New York, 1956), p. 37.
34. For an up-to-date account, see Ernan McMullin, 'Life and intelligence far from Earth: formulating theological issues', in Steven Dick (ed.), *Many Worlds* (Templeton Foundation Press, West Conshohocken, Pa., 2000), pp. 151–75.
35. www.davidbrin.com/shouldSETItransmit.html
36. www.Crichton-official.com
37. George Basalla, *Civilized Life in the Universe: Scientists on Intelligent Extraterrestrials* (Oxford University Press, Oxford, 2006).
38. Margaret Wertheim, *The Pearly Gates of Cyberspace* (Norton, New York, 2000), p. 132.
39. Stephen Baxter has produced a useful compilation of science fiction in relation to SETI and spirituality, 'Imagining the alien: the portrayal of extraterrestrial intelligence in SETI and science fiction', www.stephen-baxter.com.

10. Who Speaks for Earth?

1. http://www.davidbrin.com/SETIsearch.html
2. David Whitehouse, 'Meet the neighbours: Is the search for aliens such a good idea?', *Independent*, 25 June 2007.
3. As far as I know, no powerful laser pulses have been directed into space.
4. John Billingham, Michael Michaud and Jill Tarter, 'The declaration of principles for activities following the detection of extraterrestrial intelligence', in *Bioastronomy: The Search for Extraterrestial Life – The Exploration Broadens*, Proceedings of the Third International Symposium on Bioastronomy, Val Cenis, Savoie, France, 18–23 June 1990 (Springer, Heidelberg, 1991).
5. My wife disagrees; she is curious to know the physical form of any alien beings.
6. Douglas Vakoch, the Director of Interstellar Message Composition at the SETI Institute, has another criticism. He believes that all the messages so far composed paint an implausibly positive picture of humanity, emphasizing cooperation, artistic sensitivity and technological skill. Missing is any mention of the dark side of human nature, the wars, the planetary despoliation, the greed. The messages reflect our finest aspirations rather than the present

reality. See www.space.com/searchforlife/080410-SETI-shadow-ourselves.html.

7. See John Barrow, *The Artful Universe* (Oxford University Press, London and New York, 1995).

8. For a review, see Douglas Hofstadter, *Gödel, Escher, Bach: An Eternal Golden Braid* (Harvester Press, Lewes, 1979).

9. David Brin, 'Shouting at the cosmos', http://www.davidbrin.com/shouldSETItransmit.html.

10. I am grateful to Chris McKay for this observation.

Index